ARISTOTLE A TO ZOOS

A Philosophical Dictionary of Biology

PETER MEDAWAR

JEAN MEDAWAR

Oxford University Press

1985

Oxford University Press, Walton Street, Oxford OX2 6DP

London New York Toronto
Delhi Bombay Calcutta Madras Karachi
Kuala Lumpur Singapore Hong Kong Tokyo
Nairobi Dar es Salaam Cape Town
Melbourne Auckland

and associated companies in
Beirut Berlin Ibadan Mexico City Nicosia

Oxford is a trade mark of Oxford University Press

First published in the USA 1983 by Harvard University Press
First published in the UK 1984 by Weidenfeld & Nicolson
First issued as an Oxford University Press paperback 1985

British Library Cataloguing in Publication Data
Medawar, P. B.
Aristotle to zoos.
1. Biology—Dictionaries
I. Title   II. Medawar, J. S.
574'.03'21   QH302.5
ISBN 0-19-283043-0

Library of Congress Cataloging in Publication Data
Medawar, P. B. (Peter Brian), 1915–
Aristotle to zoos.
(Oxford paperbacks)
Reprint. Originally published: Cambridge, Mass.:
Harvard University Press, 1983.
Includes index.
1. Biology—Dictionaries.   I. Medawar, J. S.   II. Title.
[QH302.5.M4 1985]   574'.03'21   84-16529
ISBN 0-19-283043-0 (pbk.)

Printed in Great Britain by
Richard Clay (The Chaucer Press) Ltd
Bungay, Suffolk

Preface

Harvard University Press, with Peter Gay's polished 1962 translation of Voltaire's *Philosophical Dictionary* as bait, invited us to embark upon a "philosophical dictionary of biology"; but where Voltaire had taken the whole world for his canvas, we were to confine ourselves to biological topics—retaining Voltaire's format and if possible some of his esprit. We hope our *Aristotle to Zoos* title does not promise, by its A to Z format, a completeness of treatment to which we did not aspire.

This work is philosophical only in the cozy domestic sense of being leisurely, relaxed, and reflective. There is no philosophy in it of a kind that could be thought to exemplify the *Naturphilosophie* so popular in Germany in the nineteenth century—a form of scientific belles lettres with a truly dismal track record for making sense of the puzzles in which biology abounds.

It was to *Naturphilosophie* that we owed the supposed existence of the very simplest and most remotely primeval of all organisms—the *Monera*—consisting only, as the prevailing nature-philosophy required, of naked protoplasm; it was to nature-philosophy, too, that we owed the conception that the skull of vertebrates is formed essentially by the fusion and adaptive remodeling of the anteriormost vertebrae. What a shame that Thomas Henry Huxley, whose expert hatchet job on this theory of the skull was the subject of a famous Croonian Lecture at the Royal Society in 1858, should have been so far taken in by Ernst Haeckel's *Monera* as to have discovered a specimen of the group in a deep Atlantic dredging.

The spirit of *Naturphilosophie,* moreover, shines out of the misconception that cancer cells are essentially embryonic cells which, having escaped the discipline of differentiation, persist anomalously into adult life, later taking wing, so to speak, as malignant growths. All experienced biologists will share our anxiety lest the "philosophical" element in our dictionary should be construed in this deeply erroneous sense.

On the other hand, this is not a reference book and not a dictionary of explanations and definitions. There is no need for such a work so long as the *Penguin Dictionary of Biology,* a little masterpiece of expert knowledge and skillful writing, is still in print. And it has reached its tenth edition in 1983. We include a number of definitions, it is true, if only to make an evolutionary connection with a dictionary of a sort more ordinary than Voltaire's, but these definitions are grouped under the subjects upon which they bear. Personal taste turns us from the major innovation of the *Encyclopaedia Britannica* of 1771: the inclusion of very lengthy general articles dealing with whole sciences, to which seekers after particular information are repeatedly referred. We feel that anyone who, curious or fearful, looks up spina bifida wants to be told what it is and does not wish to be referred to a treatise on embryology.

Perhaps the most famous passage in Voltaire's dictionary comes in the chapter "Tout est bien," which contains the black theodicy beginning "Either God wishes to expunge the evil from this world and cannot; or he can and does not wish to; or he neither wishes to nor can." The entire passage was quoted by Voltaire from a treatise concerning the wrath of God by the fourth-century church father Lactantius, who professed to have had it from Epicurus. Voltaire's contemptuous dismissal of Lactantius' attempt to controvert the implications of this passage makes us think that Voltaire would have had little patience with the kind of theological humbug brought to the attention of the world by the Reverend Lord Bridgewater, whose persuasions we shall consider in our text.

If the present book is not for reference and not for looking up things, what—and whom—is it for? It is for browsing. For the best results the reader must abstain resolutely from learning anything he or she does not want to know and must exercise at all times the reader's liberating privilege—skipping.

Among the educated audience we envisage might be biologists, sociologists, psychologists, and other members of that increasing population of reflective people who see in biology the science most relevant to the understanding and melioration of the human condition.

PREFACE

We are grateful to the following for permission to reproduce the figures and photographs that appear in the text: to the Science Museum, London, for the figures illustrating the three-dimensional geometric realization of a number of algebraic functions; to Clarendon Press for the illustration of the Gompertz function and its principal derivatives; to the Royal Society of London for the illustration of how curves of growth may be confounded by curves of distribution; and to Cambridge University Press for the figures from Sir D'Arcy Wentworth Thompson's *On Growth and Form.*

It is a pleasure also to record our special indebtedness to our literary assistant, Mrs. Joy Heys, for her assistance in preparing the manuscript and to Vivian Wheeler, senior editor at Harvard University Press, for her scrupulous editing and ordering of our manuscript.

P.B.M.

J.S.M.

CONTENTS

CONTENTS

CONTENTS

CONTENTS

ADAPTATION

This word is used in biology in several different senses, the contexts of which are so different that they are not likely to give rise to misunderstandings.

Adaptation may refer to the transformation occurring in the course of evolution that fits an organism to its habit or habitat (the wings of birds adapt them to flight and the snouts of anteaters adapt them to eating insects); in a sense all evolution is adaptation.

The word also refers to a change in the life of an individual organism that enables it to cope better with the hazards to which its environment may expose it: the formation of corns or calluses where a shoe rubs is one such an adaptation, and the formation of an antibody against an infectious organism is another; again, at low oxygen tensions—as at high altitudes—the number of red blood corpuscles increases, and repeated exertions enlarge the muscles that perform them. Adaptive changes of this kind are not imprinted on the genome and therefore are not heritable.

Adaptation in the context of sensation refers to the fact that a prolonged and uniform sensory stimulus eventually ceases to give rise to a sensory message; thus someone who goes into a room containing a bowl of roses may smell them at first, but then become unaware of them. Once this process of sensory adaptation has occurred, no effort of attention can call the smell to mind, so this is quite unlike our unawareness of the ticking of a clock, which an effort of attention *can* at once recall to mind. It is a popular fallacy that chewing gum regains its flavor if removed from the mouth and parked, say, under a chair. What is regained is not the flavor but the ability to taste the flavor as sensory adaptation wears off.

ADRENAL GLAND

The adrenal is an endocrine gland—one that liberates its secretory product directly into the blood or lymph, instead of into a duct which transports it elsewhere. It is in some ways an exemplary gland, for it illustrates especially clearly a number of functional, evolutionary, and developmental characteristics found in endocrine glands generally.

The adrenal is a paired organ, somewhat diffuse and untidy looking in lower vertebrates, but in mammals a compact, paired organ lying at the anterior poles of the two kidneys.

The adrenal gland is composite in origin, as are many other endocrine glands. Indeed, the two components of the gland in mammals are entirely distinct in origin and function. The central core, or *medulla,* is essentially a modified ganglion of the sympathetic nervous system. The adrenal medulla secretes adrenaline (epinephrine) and noradrenaline, the effects of which are, understandably, the same as those brought about by general sympathetic stimulation: cardiac output is increased and the blood vessels of the brain, retina, and kidney are contracted, while the blood vessels in skeletal muscle tend to dilate. The psychological effects are familiar to people who have had injections of local anesthetics for dental operations. To discourage local bleeding, such anesthetics often incorporate adrenaline in quantity sufficient to give the patient feelings of apprehension, restlessness, and unease superimposed upon feelings of the kind so readily stimulated by a dental operation, or the prospect of one.

The outer shell of the adrenal gland—the adrenal *cortex*—may arise in development from the first set of kidney tubules to be formed in life, the so-called pronephros. The cortex, which unlike the medulla is essential for life, is concerned with the regulation of minerals in the body, especially sodium salts. Removal of the adrenal cortex—or its destruction as a consequence of infection or injury—causes rapid salt depletion. The effects can be fatal unless the victim receives injections of the natural secretions of the gland (including aldosterone, corticosterone, hydrocortisone—or cortisol, and other related substances) or of substances exercising the same action. Because the symptoms of withdrawal of its secretions include a general enfeeblement and diminished resistance to stress,

2

it is generally assumed that the cortisone-like secretions of the adrenal cortex are essential for the normal functioning of the body, particularly the integrity of the skeletal and supportive tissues and resistance to stress.

That part of the cortex which is responsible for the secretion of cortisone-like substances is under the control of a secretion of the pituitary gland named, in accordance with the usual convention, adrenocorticotrophin (ACTH for short). When the adrenal cortex is in working order, injection of ACTH to increase its output is preferable to injection of the cortical hormones themselves. The steroids sometimes referred to as "anabolic" promote anabolism and in high doses may increase muscle volume considerably, a property taken advantage of by athletes such as weight lifters, especially in countries which regard victory in competitive games as evidence of the superiority of their political order.

AGE DISTRIBUTION

A population has a structure not merely in space (its geographic distribution) but also in time, for two populations may have very different compositions with respect to the ages of their members.

The age distribution is shaped by the prevailing death rates and fertility rates; only if the same regimen of both has been in force for many generations does the age distribution adopt a stable character, one which would be reestablished after temporary fluctuations or disturbances due to migrations of population or to wars and the like.

The age composition of a population is profoundly important politically because it is the principal factor that determines the social burdens imposed by education, if state aided, and old-age pensions, if granted. It may also affect the political temper or voting behavior of the population because of the widespread but probably erroneous view that the young, being hotheaded, will favor liberal or radical policies and vote accordingly, whereas the older, having become set in their ways, will be in favor of the status quo. As Georges Clemenceau, Premier of France (1917–1920), once said, "A man who is not a socialist at twenty has no heart; a man who is still a socialist at forty has no head." We have little doubt that sociological research would find the grounds to prove both of these beliefs illusory.

It can be very misleading to compare two populations with respect to, say, fertility or mortality if their age distributions are widely different. Mortality from cancer or any other disease with an age-related incidence in different countries can be compared only if the incidence of mortality in the two countries is referred to a certain standard age distribution; for mortality in one country may be higher than in another simply because it has a higher proportion of older people. It was his disregard of considerations such as these that led David Livingstone to formulate the well-known half-truth that "cancer is a disease of civilization." A population may increase in number by the accumulation of members beyond reproductive age, while at the same time its power to reproduce itself may be in decline.

AGGRESSIVE INSTINCT

People who speak of "the aggressive instinct" (few professional biologists are among them) are usually aggrieved and resentful when it is pointed out to them that there is no such thing: that is to say, that there is no kind of drive or appetitive behavior that is assuaged or discharged by aggressive performances. There are, however, aggressive elements in a variety of kinds of instinctive behavior—in defense of territory, for example, in sexual rivalry, and in competition for food. Some of this aggression is known to ethologists to be bluff—to be a kind of warning demonstration or saber rattle that may achieve without risk to the combatants all that real aggression would do. There is no substance whatsoever to the argument that, because "man is an aggressive animal" and "you can't change human nature," wars between nations are the inevitable consequence of aggressive instincts. Educated people do not have to refer to ethological papers before rebutting these trite propositions. Indeed, such phrases are now almost universally taken to mark an especially low level in conversation or discussion.

AGING

Although taken literally "aging" merely means growing older, it is seldom used without the connotation often distinguished in professional speech by use of the term *senescence*—the progressive deterioration of bodily faculties and performances that accompanies growing older. Attempts have been made from time to time to represent senescence as a pathological process, yet its character and pattern of onset in human beings and domestic animals are such as to leave one in very little doubt that it is in reality a natural process—and of a piece with development, of which it is the closing episode.

When does senescence begin? Examination of our own condition tempts us to answer, "after a period of growth and maturation," but in an arresting series of articles published in *Popular Science Monthly* (and republished as *The Problem of Age, Growth and Death* [London, 1908]) the American anatomist Charles Minot gave a very different answer: aging, he declared, begins at birth. For if we take the specific growth rate—measuring the power of living tissue to reproduce itself at the rate it was itself formed—as the best measure of "vitality," then it can be said with confidence that senescence does begin at birth because the specific growth rate falls from birth on.

Minot's initiative was the consideration that prompted us to describe as Minot's Law the second of the two laws of growth we propound in the entry on GROWTH, LAWS OF BIOLOGICAL; this law affirms that *the specific acceleration of growth is always negative.* There is, however, an element of preciosity about saying that aging begins at birth; with less violence to conventional judgment it could be said that bodily deterioration begins immediately after the actuarial prime of life, the brief epoch of life in which the force of mortality is at its minimum and our likelihood of living another month or day or year or minute is higher than it is at any stage before or after.

This remark is a suitable cue for introducing the actuarial measurement of senescence, that which is almost universally adopted by students of aging. According to this scheme of measurement, senescence is conceived as that change in bodily structures, faculties, or energies which increases the likelihood of dying from accidental causes of random incidence, for after the actuarial prime of life we become ever more vulnerable. If people died only from accidental causes of random incidence and aging did not affect their vulnerability, then the force of mortality would be constant, but in reality it rises from the actuarial prime until the end of life.

Although this is an apt and sensible scheme of measurement, it is not foolproof: it would be pedantry, for example, not to describe the menopause in menstruating animals as an episode in the process of senescence; yet the menopause does not increase a woman's vulnerability to mortal hazards, and to people of sanguine temperament it could be taken as an outward sign that one class of hazards has been outlived, namely those that have to do with pregnancy and childbirth. Likewise the graying of hair, universally accepted as a manifestation of aging, is not accompanied by an increase in vulnerability. This of course might be taken to import that the graying of hair is not "really" a senescent change, but only a convenient outer manifestation of the contribution made to bodily deterioration by the accumulation of minute accidents and traumata such as radiation hits or physical injuries that destroy the melanocytes that feed their coloring matter into hair.

Still, it is not very easy to distinguish between those senescent changes of the kind sometimes described as "programmed" (that is, developmentally "laid on") and those that represent the cumulative sum of minute and individually trivial accidents or malfunctionings. Indeed, according to one theory of aging (the one that is particularly associated with the name of Leslie Orgel), such a distinction cannot be validly drawn. For senescence itself is the consequence of an accumulation of errors of transcription or translation from the genetic message specifying the synthesis of enzymes in the course of synthetic and vital processes generally.

There is, however, one sense in which it is reasonable to describe senescence as a pathological process: there is no known reason why it should *have* to happen. Biologists no longer accept (and indeed dismiss as Panglossism) August Weismann's notion that senescence occurs for the benefit of the species, to remove the spent or worn-out animals that Weismann somewhat unrealistically thought were cluttering up the environment and depriving numerous youngsters of food and lebensraum. What kind of phenomenon, anyway, would make senescence an obligative process? One possibility would be a determinate life span of cell lineages—a limitation of the number of successive cell divisions through which a clone of cells could pass. Weismann, curiously, believed that such a limit existed. Although he professed not to know why it should be so, a constitutional lack of quantitative judgment made him suppose that a cell might be confined to one thousand or ten thousand successive divisions. Whereas he was not wholly wrong in principle, quantitatively he was very wide of the mark. Although we all at one time believed that cells could be propagated indefinitely in tissue culture, we now know that normal cells have a finite lifetime and cannot be carried through an indefinite number of cell divisions under any circumstances.

The real reason why the flagrantly unadaptive process of senescence takes place is that from an evolutionary point of view it makes little difference whether it does or not. In the entry on HUNTINGTON'S CHOREA (a genetic disease of dominant determination, which makes itself manifest relatively late in life) we point out that most of its future victims could have had a family of ordinary size before the disease appeared. This is probably true of all manifestations of senescence prominent or debilitating enough to be attributed to aging: they occur during the postreproductive period of life, after people or other organisms have taken whatever share of the ancestry of future generations could be expected of them. The force of natural selection is greatly attenuated; considered as a nonadaptive character, then, senescence cannot be opposed by the common selective forces and phased out of the genome of any species.

This circumstance led to the advancement of the theory of aging especially associated with the names of P. B. Medawar and G. C. Williams, namely that aging is the consequence of the accumulation in the genome of maleficent genes, the ill effects of which do not become apparent until late enough in life for them to be virtually out of reach of natural selection. If we look at Huntington's chorea and a number of other diseases of late onset (such as familial intestinal polyposis, often leading to cancer) as cases in point, it is undeniable that the theory has some substance. It is a corollary of the theory—and was indeed one of the considerations that led to its being propounded—that senescence is essentially an artifact of domestication. This does not of course mean that domestication is the cause of senescence, merely that it is the unnatural state of affairs that allows senescence to become manifest by withholding from the organism the action of many of the agencies that in real life might easily have been fatal (privation, starvation, exhaustion, or infection). Indeed, contrary to popular superstition, aged animals are something of a rarity in nature, where a savage exaction of mortality is the rule.

Senescence, anyway, cannot be regarded as an epiphenomenon of life: it is something we must live with. On the other hand, senescence is a physical change and there is no reason in principle why some means should not be devised to meliorate it. One of the great strengths of the ethos and scheme of manners implicit in puritanism is that judging from experience with laboratory animals, temperance (especially in calorie consumption) is a very important—perhaps the most important—agency for the prolongation of life. Several investigators have recorded that systemic administration of high doses of industrial antoxidants brings about a significant prolongation of life. This observation has not been welcomed as rapturously as might at first be thought because of the uneasy suspicion that the antoxidants, which are somewhat toxic, could act by diminishing appetite, thus reducing calorie consumption to the degree that tends to favor longevity. The mention of high doses of antoxidants systemically administered turns the thoughts toward oral consumption of high doses of ascorbic acid (vitamin C), which, too, has its advocates—the two chief of whom, lively octogenarians, are alive at the time of writing.

ALBINISM

This is a disorder of the pigmentary system in which, with no diminution in cell number or change in disposition, there is an extreme attenuation of pigmentary function in the cells potentially capable of making pigment—the *melanocytes*. Unmasked by black pigment, the retinas of albino people and animals, being very highly vascular, are bright pink. Without the protection of pigment, the retinal sensory cells are easily damaged, and albinos generally suffer severe defects of vision.

Albinism, or rather the dilution of pigment of which it represents the extreme form, is genetically controlled and is to be found in many domestic animals such as mice, rats, guinea pigs, horses, and human beings. In some cultures albino human beings have been regarded with superstitious awe, and white steeds much admired.

In so-called albino guinea pigs and Himalayan rabbits their condition, referred to as albinism, is in reality an extreme dilution of pigmentation. These animals have the property that their points (paws, tips of ears and noses) darken in cool weather for reasons not yet understood.

Because they look alike, albino rats and mice are often assumed to be genetically uniform; in reality, albinism may conceal what would otherwise be flagrant genetic diversity. This is a familiar trap for inexperienced experimental biologists and psychologists. The latter, for instance, have often taken Wistar rats to be as uniform in genetic makeup as they are in color.

ALLERGY

An allergy is a miscarriage of the immunologic process in which something that begins as a protective device has deleterious consequences that range in severity from itching to sudden death.

The allergies are protean in their diversity, and not all their manifestations or all the agents that can arouse them are yet known. Pretty much any nonself substance that can gain access to an immunologic response center *may* be an allergen—that is, an antigen that is a prime mover in an allergic response. The best known and most tiresome antigens are the pollens that arouse hay fever and asthma, the fungal spores that give rise to farmer's lung, the mites in house dust whose excrement arouses an especially intractable form of asthma, the industrial chemicals that are forever arousing allergic dermatitis, and the many antigenic foodstuffs such as eggwhite and shellfish which are notorious for causing gastric pains and urticaria.

Other grave offenders are new medications or their vehicles, substances such as antibiotics and especially the horse serums that contain antibodies to bacterial toxins—antidiphtheric serum and antitetanus serum, for example. (It has been cynically said that, considered as an allergen, horse serum has caused more deaths than have ever been prevented by therapeutic action of the antibodies of which it is the vehicle.) The liberation of histamine by a certain kind of white blood corpuscle during the interaction of antigen and antibody is responsible for the physical distress of *anaphylactic shock;* other such pharmacologically active substances arouse allergies of the so-called Arthus type, triggered by the meeting of antigen and antibody within the tissues. (Farmer's lung is a reaction of the Arthus type.)

A theoretically important and medically useful form of allergy is the delayed-type hypersensitivity reaction. It is described as "delayed" to distinguish it from the rapidly developed wheal-and-flare, which is caused by the injection into the skin of test quantities of antigens that arouse allergies of the asthma/urticaria type. A hypersensitivity reaction of the delayed type, on the other hand, is characteristically aroused by the introduction into the tissue of antigens of the kind that arouse cell-mediated immunity. Over a period of five to twenty-four hours swelling and reddening develop, accompanied by some pain or itchiness and—as revealed under microscopy—a congregation of white blood corpuscles (especially lymphocytes) within the tissues near the point of injection. The tuberculin reaction is an example: it is the basis of the *Mantoux* test, in which such a local skin reaction is aroused by very small quantities of extracts of tubercle bacilli in people who have or have had an active tubercular infection.

Most allergies are not so potentially useful as the Mantoux test; on the contrary, they are a significant source of malaise, discomfort, or distress. Antihistamine drugs and adrenal cortical steroids are widely used for the symptomatic treatment of allergies but, as ever, prophylaxis is best—the avoidance wherever possible (and very often, of course, it is not possible) of allergens such as pollen, fungal spores, house dust, and sensitizing chemicals.

ALLOMETRIC GROWTH

It is the merest truism that growing up is not merely a matter of symmetrical enlargement, that the adult is not merely the youngster writ large. On the contrary, organisms change in shape as they develop, and inasmuch as this is an orderly and regular process we should expect to find some orderly and perhaps simple relationship between the growth rate of one part of the body and the growth rate of the whole or of another part.

The "law" of allometry first proposed was that in growing up the specific growth rate of one bodily part bears a constant ratio to the specific growth rate of another part. If x and y designate two parts and stand also for their sizes, the specific growth rates of the parts are respectively $(1/x)$ $(dx/dt)=(d \log_e x)/dt$ and $(1/y)$ $(dy/dt)=(d \log_e y)/dt$ so that their ratio is (x/y) (dy/dx), or $(d \log_e y)/(d \log_e x)$. In the original formulation by Julian Huxley this ratio was said to be constant, so that if the logarithms of the two variable quantities were plotted against each other they would define a straight line according to the linear equation $\log y = \log b + a \log x$.

This law is not now thought to be universally valid, though there are some striking conformities to the rule. From the character of the equation it should be noted that $a = 1$ when x and y grow strictly in proportion to each other, and that y is gaining on x or alternatively losing ground as a is greater or less than unity.

For a more general treatment of the problem of change of shape in development, the reader is referred to our entry on TRANSFORMATIONS. See also the discussion of GROWTH, LAWS OF BIOLOGICAL.

ALTRUISM

Altruistic behavior is behavior that brings advantage to others than those who perform it. The connotation of disadvantage to the performer is not an essential part of the meaning, but is rather a special property called in evidence to show that behavior is genuinely altruistic and not motivated by covert self-interest.

Altruism raises some very nice problems in the genetic theory of evolution by selection—an entirely selfish process that depends upon a gene's enjoying a net reproductive advantage over its alternative forms; but how can any gene have such an advantage if it disadvantages its possessor? One answer, which plays a crucially important part in the theory of sociobiology, is *kin selection.*

Consider parental care, for example. To use words that might have come naturally to the late P. G. Wodehouse, a hen that becomes broody is particularly vulnerable to a fox or other predator that creeps up behind it wielding a short length of lead piping; how then can the genes that provide for socially advantageous behavior of this kind become established in the stock, as it is necessary and desirable that they should? The answer is that a parent that protects its young is at the same time protecting those of its genes that are represented in the young. J. B. S. Haldane put it more pointedly in an aside recollected by John Maynard Smith: he said he was willing to lay down his life for two brothers or eight cousins. The genes promoting this highest form of altruistic behavior ("Greater love hath no man . . .") would have prevailed in the population and brought net reproductive advantage to their possessor's genes even if their possessor had lost his life. In such a case the advantages or disadvantages can be fairly easily quantified, but some reports of altruistic behavior should be regarded with extreme skepticism, the most notorious being the widespread belief that in order to reduce their numbers and thus promote the welfare of the species as a whole, lemmings from time to time commit suicide in large numbers by throwing themselves off a cliff to drown. "If you believe that, you can believe anything," as the Duke of Wellington is said to have remarked to a lady who accosted him with the words "Mr. Smith, I believe?"

The theoretic importance of kin selection is that it lends a degree of credibility to the kind of evolutionary speculation embodied in sociobiology.

AMINO ACIDS

An amino acid is the building block or monomer from which the protein molecule is constructed. It may be thought of as a compound formed by substituting an amino (NH_2) group for one of the hydrogen atoms on the carbon chain of a fatty acid. Thus the substitution of an NH_2 group for one of the hydrogens in acetic acid CH_3COOH, the characteristic acid of vinegar, yields the simplest amino acid, *glycine:* CH_2NH_2COOH. An amino acid (there are twenty different kinds) is effectively acidic or basic in accordance with whether carboxyl ($COOH$) or amino groups predominate in the molecule.

AMPHIOXUS

This zoologically famous animal, of which the agreed generic name is now *Branchiostoma* (lancelet), owes its importance to being the modern representative of the subphylum *Cephalochordata,* which was surely ancestral to the vertebrates. Displaying in almost diagrammatic form the defining characteristics of the group Chordata, amphioxus is an inch or two long and, like a modern automobile, is pointed at both ends (hence its older name, the prefix *oxy* having the connotation of sharpness or acidity).

Although it can swim, amphioxus normally lives in shallow seawater on a sandy or gravelly substratum. Like its ancestor the sea squirt, it is a filter feeder, living on small particles suspended in the water current that passes through its mouth into a spacious pharynx, the antechamber to the gut. The pharynx is perforated by numerous slits through which the seawater escapes, whereas solid matter is trapped on the walls of the pharynx made sticky by a mucous secretion from the median ventral groove, the *endostyle* that is the precursor in vertebrates of the thyroid gland. The presence of a large perforated pharynx associated with the habit of filter feeding is characteristic of chordates.

Another chordate feature is the elastic skeletal rod, the *notochord,* that stretches from one end of the body to the other. The musculature consists of segmentally arranged muscle blocks also extending the length of the body. There are no limbs. The central nervous system is of the usual chordate pattern: a median dorsal tube sends motor nerve fibers to the segmental muscle blocks and receives sensory fibers that pass between the muscle blocks and enter the tube dorsally. The blood system, equally typical, is a median ventral contractile vessel propelling blood forward and up the sides of the pharynx in the gill bars that separate the gill slits. Circulation up the gill bars is assisted by a small "branchial" accessory heart at the base of each bar.

One system is altogether atypical of chordates. Anatomists were greatly surprised when the excretory system was shown to consist of little clusters of excretory cells, "flame cells" such as are found in several invertebrates. Aside from this one uncalled-for departure from the chordate canon, amphioxus answers so very well to what is expected of an organism ancestral to vertebrates that it would have been said to exist (or to have existed) even if it had not been discovered and recognized.

So much for the rich relatives of amphioxus. Its poor relatives are revealed by the possession of a pharynx and the filter-feeding habit, and even more so by its development; for the earliest developmental stages of amphioxus are almost identical to those of the sea squirt. There is another affinity too: the larva of amphioxus is curiously asymmetrical, the mouth starting high up on one side of the front end of the body and moving into its definitive position only by a complex rotatory movement of the entire branchial basket in relation to the notochord and muscle blocks. It has been suggested that this rotatory movement is homologous with the rotation of the pharynx in the embryonic sea squirt, a strange metamorphosis that precedes its attachment to the substratum and causes mouth and anus to occupy their definitive positions in the adult sea squirt.

ANABIOSIS

Now obsolete, this term refers to the state of suspended animation characteristic of tissues preserved by extreme cold.

The idea of preserving whole organisms in this way occurred first to John Hunter (1728–1793). "Of the Heat of Animals" describes how he tried to freeze two carp, hoping that "it might be possible to prolong life to any period by freezing a person . . . I thought that if a man would give up the last ten years of his life to this kind of alternate oblivion and action, it might be prolonged to a thousand years; and by getting himself thawed every hundred years, he might learn what had happened during his frozen condition. Like other schemers, I thought I should make my fortune by it; but this experiment undeceived me" (from *The Works of John Hunter, FRS,* ed. J. F. Palmer [London, 1835], vol. 1, p. 284).

ANENCEPHALY

Gross reduction in the size of the cerebral hemispheres so often accompanies spina bifida that they are frequently classified together as a single syndrome, ASB, standing for anencephaly/

spina bifida. It is a developmental aberration to which the structures of the embryonic axis are especially vulnerable. ASB is not in any ordinary sense a hereditary affliction, but for unknown, possibly immunologic, reasons it does occur more frequently in women who already have a history of spontaneous abortions. One embryologist has suggested that its higher frequency among Roman Catholics may be attributed to the use of the rhythm method of fertility control, which tends to result in late fertilization and thus in an abnormally long delay in activation of an egg cell by the ordinary process of fertilization. There are pronounced variations from place to place and from population to population, the incidence being notably high in Northern Ireland and rather low in Africa south of the Sahara, though epidemiologists are always on their guard against such comparisons because of the variation from place to place in the reliability of ascertainment. Congenital abnormalities generally offer no comfort to racists because some are more frequent, others less so, in black than in comparable white populations.

Anencephaly is an irremedial and almost invariably fatal birth defect, though surprisingly the degree of anencephaly associated sometimes with hydrocephalus is compatible with some cerebral function made possible by the amazing functional plasticity of the brain.

ANIMALS AND HUMAN OBLIGATIONS

The word "animal" is not used in King James's Bible, but Jehovah's rulings on the relationship between man and beast are clear enough. Human beings were to have "dominion over the fish of the sea and over the fowl of the air, and over cattle and every creeping thing" (Gen. 1:xxvii). A later verse goes on to reassure man of "dominion over the fowl of the air and over every living thing that moveth upon the earth." Plants rank low in the hierarchy, for to all the animals mentioned "I have given every green herb for meat" (Gen. 1:xxx).

The divine remit did not encourage compassion for animals or any attempt to understand them as animate creatures. The prevailing indifference is understandable, inasmuch as peasants whose own lives were very hard would probably not have thought the beasts so much worse off than themselves.

The changed attitude toward animals may be thought of as a late by-product of the Romantic revival, signaled perhaps by the foundation in England in 1824 of a Society for the Prevention of Cruelty to Animals and perhaps by the publication of a skillfully written and widely popular "autobiography" of a horse—*Black Beauty* by Anna Sewell, published in 1877. The subsequent growth of literacy encouraged proliferation of a nursery literature in which rabbits, cats, foxes, lions, tigers, elephants, and bears all had speaking parts.

To children this seemed quite natural, and they took it in their stride. Our younger daughter, Louise, as a five-year-old was taken to a zoo; there she saw a large parrot high up in its wire cage. "Hallo!" she said civilly, having been properly brought up. The parrot climbed slowly down from the very top of the cage: beak, claw, beak, claw until it was at the child's eye level. It then thrust its head forward, said " 'allo!" and climbed very slowly back up, leaving Louise elated at the thought that she was now on speaking terms with a parrot.

Our modern concern over the welfare of animals seems to have originated in the general increase of sensibility that followed the Second World War. Perhaps also a factor was the propaganda and persuasive literature of that era, urging upon human beings an awareness of the duties that go with their trusteeship of animate nature. One beneficial result has been the establishment of game preserves and conservation land.

This new awareness has been accompanied by a movement strongly critical of a number of morally ambiguous but socially accepted practices that may disadvantage animals unreasonably.

The first of these practices is experimentation on animals, justified only by the benefits it is rightly believed to confer upon human beings. Those whose affection for animals is so extreme as to encourage an indifference to human welfare that borders on misanthropy advocate a total halt to experimentation on animals. They undermine their own case and lose the confidence of the public by the recklessly inaccurate assertion that experimentation on animals has brought no benefits at all to mankind—a feat of doublethink made possible only by attaching more weight to the encephalitides very rarely caused by whooping cough vaccine than to the combined benefactions of insulin, penicillin, polio vaccine, and transplantation surgery. It is a pity they adopt this attitude, because it weakens and may destroy altogether the valid case for legal regulation of experimentation on animals.

If Samuel Johnson were alive today, he would condone the use of animals for the safety testing of putatively therapeutic drugs, but would scathingly denounce the use of animals to test beauty preparations and cosmetics. No face, he would argue, is so pretty as to justify paining an animal. Furthermore, he would surely insist—as British legislation now does insist—that experimentation on animals should be executed only by those whose training and declared intentions make clear to a panel of impartial judges that they are technically competent. Under no circumstances should experiments ever be performed merely to gratify curiosity or to add to the treasure-house of human knowledge. All experiments must have a specific and declared medical or scientific purpose, always such as to make possible or easier the fulfillment of medical ambitions. All this involves the certification of experimentalists and a continuing review of their activities, supplemented—if thought fit—by unannounced site visits. And, of course, a bureaucratic apparatus that must be large enough, tough enough, and sufficiently well informed to have a cutting edge.

These desiderata are provided for by the regulations currently in force in Great Britain. Most northern European countries exercise some degree of regulation of experimentation on animals (see *New Perspectives in Animal Experimentation,* ed. David Sperlinger [New York, 1981]). The United States, currently impatient with governmental interference, especially by a restrictive bureaucratic machinery, seems to be comparatively backward in these matters, something unexpected in a country which has so often led the way in legislation for the public good.

Although the supervision of animal experiments sounds as if it would be yet another uncalled-for intrusion of officialdom into the lives of respectable private citizens, no competent and humane experimenter should have reason to object to certification or to official supervision or to unannounced inspection visits. Research would be gravely impeded by enforcement of a requirement that each single experiment (such as the injection of one mouse with penicillin) be specifically authorized by high dignitaries of church and state; for the bureaucratic process is well known to feed upon itself and to multiply far beyond the minimum necessary to transact its business. It should be emphasized, though, that the officers of licensing authorities (most of whom do their work conscientiously and expeditiously) can be a source of useful advice.

Once they have official permission, experimenters can make a start at meeting the standards put forward by W. M. S. Russell and R. L. Burch in *The Principles of Humane Experimental Technique* (London, 1959). Their three Rs are Reduction of the number of animals used, Refinement of experimental procedures, and Replacement of living systems by inanimate ones such as tissue culture.

Blood sports constitute another socially accepted practice that disadvantages creatures lower than man. Killing animals for fun belongs to a tradition so old and so closely associated with what are thought of as manly virtues that it is not counted a malpractice even when some rich and doubtless overweight mechanized vacationer off the coast of Florida whiles away his time by assassinating a shark or some other magnificent sea beast.

It would benefit hunters as well as hunted if sportsmen realized that their considerable exertions to kill animals for fun may eventually coarsen appearance, speech, and manners. It is no wonder that in *A Woman of No Importance* (1893) Oscar Wilde said of "the English country gentleman galloping after a fox" that it was "the unspeakable in full pursuit of the uneatable."

It is not generally known that among the folk most guilty of cruelty to animals—anyhow, to dogs—are professed animal lovers themselves. We have in mind the practice, rapidly dying out under pressure from the veterinary profession, of breeding dogs for perpetuation of the exhibition points that count high in shows. Many of these breeders' points are frank abnormalities. Consider, for example, the sadly crumpled face of the bulldog with its deeply folded and fissured skin, predisposed to dermatitis. The bulldog's difficulty in breathing is due to the skeletal abnormalities of its head, putting one in mind of the mutant abnormality known as *achondroplasia;* the prized "gay" or curly tail is an abnormality of the development of the vertebral column. To these we may add the admired show stance that is made possible by congenital dislocation of the hip. Again, some dogs are bred for the cultivation of a mop of hair that falls over the forehead and causes chronic irritation of the eyeball. Other dogs have hair so thick that in warm weather they have the utmost difficulty in keeping cool by excreting heat and often are thermally distressed. The British Small Animals Veterinary Association classifies all such conditions as enormities and declares that breed societies and show judges are deeply mistaken to connive at their perpetuation. Vanity is at the root of these excesses, as it is also at the root of the deliberate breedings of dogs as toys—as miniatures that will just fit into a deep pocket or handbag, for example. Many of the people that condone or even applaud these malpractices arrogate to themselves the title of dog lover and are among the loudest in denouncing the use of animals in medical research.

Another misuse of animals is now arousing loud self-reproach: the treatment of animals as, in effect, manufactured products—or simply as machines to convert vegetables into meat, as is the case with battery chickens and other animals that can be subjected to mechanized husbandry. Although it is not to be supposed that the animals so treated enjoy the refined sensibilities of a lake poet or that they pine for "daylight and champain," it cannot be denied that the entire proceeding is an affront to one's sense of the fitness of things, and a moral diminishment of man. It seems doubtful that anything short of legislation will bring the practice to an end, for battery farming provides an abundant, commercially profitable source of relatively cheap first-class protein. Moreover, the enterprise has acquired, among people who rate themselves realists, the tawdry glamour associated with anything commercially advantageous, even if the advantage has been secured by somewhat questionable means.

The record of human aggression as it affects other species is a dismal one that records the extinction of roughly three hundred species in the last three hundred years, among them passenger pigeons, that affable dugongid Steller's sea cow, and the Tasmanian wolf. The Tasmanian people, too, have been exterminated (the last native Tasman died in 1876). Jared M. Diamond of the University of California has, however, warned us not to accept too readily the Arcadian legend that this is mainly the work of mechanized Western man, or that man the hunter-gatherer has always been a shrewd and prudent conserver of whom a modern ecologist could be proud.

That our ancestors have sinned does nothing to excuse the heroic scale on which modern hunters, prompted by commercial interests, have raided nature—something that is rapidly leading to the extermination of whales in spite of the best endeavors of United Nations' agencies. It is a sad note that politicians have passively acquiesced in the notion that commercial advantage excuses almost any enormity and that those who protest and actively express concern for the welfare of animals are dismissed as sentimental cranks, eccentrics, or (in British schoolboy terminology) "wets."

What can be done? Hortatory speeches are not enough and certainly it is inefficacious to call for worldwide refinement of human sensibilities in the interest of plants and animals. What we need is a case for "doing good in minute particulars." Contemplation of the abuses we have cited suggests remedial legislation. All that one asks of human sensibilities is that the issue be taken seriously enough to carry political weight. Nothing is quite so effective a catalyst of legislation as the feeling that it will bring political advantage to the person or party that initiates it. Unhappily, in addition to having no rights in common law, animals generally have no political clout whatsoever.

ANTIBIOTICS

It is fortunate for human beings that no Geneva convention forbids the prosecution of germ warfare by germs themselves; if it did, there would be no antibiotics—substances produced by some microorganisms that destroy or arrest the growth of others. Although the teleology of antibiotics is still not absolutely certain, we take the reasonable view that they are the principal agents of competition among microorganisms in the densely crowded environment of the soil (soil bacteria being a principal source of antibiotics).

The use of antibiotics for therapeutic purposes is made possible by the fact that at the nucleic acid level inheritance and information transfer are the same in microorganisms as they are in higher animals. As these processes are the most vulnerable targets for antibiotics, they may be used to kill or impede the growth of bacteria in vivo—provided, of course, that their action falls short of poisoning the higher organism that is to be cured of its infection.

The discovery of the penicillins, combined with the earlier discovery by Gerhard Domagk of the antibacterial action of sulfonamides, constitutes the most important advance in medicine during the twentieth century. It ended the tyranny over man of microorganisms, the most important selective force at work upon human beings since the beginnings of their evolution.

It is well known that penicillin was discovered by accident by Alexander Fleming. As the product of a mold, *Penicillium notatum* spread by airborne spores. It is less well known that Fleming, at one time an employee of the British Medical Research Council and a researcher during the First World War into the control of infection in war wounds, had for many years been looking for penicillin or something very like it—for a nontoxic antibacterial; that is, for an antibacterial substance that did more harm to the bacteria than to the organism they had infected. Our lack of awareness of this important psychological motivation generated a number of what are now known to be legends about the discovery of penicillin (such as the spore floating through the open window, settling on a bacterial culture plate, and causing the formation of a halo of inhibition of bacterial growth). The propagation of this legend distracted attention from the one piece of genuinely astounding good luck associated with the discovery of penicillin: that penicillin turned out to be nontoxic. For it has since been learned that antibiotics as a class are very toxic indeed; the kind of damage they do to bacteria, they do also to the organisms they parasitize. Penicillin owes its relative nontoxicity to the fact that the bacterial metabolism it interferes with is pretty well peculiar to bacteria, being relatively specific to processes involved in synthesis of the bacterial cell wall, a circumstance that enormously reduces the toxicity for higher animals.

Indeed, the discovery of penicillin and its extraction from cultures in a quantity large enough to treat mice was only the beginning of the story. Howard Florey's laboratory, in which the authors of the present work carried out their graduate studies, was a typically penurious and ill-equipped university department totally unfitted to produce penicillin on the scale necessary for clinical trials. Although a number of British firms undertook the manufacture of penicillin, it was not until American wealth and know-how was brought to bear upon the matter that penicillin was produced in adequate quantity. Florey and members of his team visited the United States and openly shared all the knowledge they had acquired about extractions carried out on a modest scale. The American firms with whom they had met patented this information, with the result that when British firms got around to manufacturing penicillin themselves, they had to pay royalties to the Americans for the privilege of doing so. The story, recounted by Florey's biographer, R. G. Macfarlane (in *Howard Florey: The Making of a Great Scientist* [Oxford, 1979]), is a thoroughly discreditable one and has taken its place among the principal case histories of "What's wrong with capitalism?" Whatever the morals, American wealth and technology did what was really important: they produced clinically adequate quantities of penicillin at a time in history when such a medicament was wanted more urgently than ever before.

Since those days, penicillin has been greatly improved. Many new antibiotics have been discovered, constituting now a veritable armory. Even some of those that are highly toxic have their uses in, for instance, the treatment of cancer.

Antibiotics of bacterial and fungal origin do not, in general, act upon viruses. INTERFERON, the subject of a separate entry, is antiviral in action and is not an antibiotic in the sense in which that term has been used in the present entry.

ANTICHOLINESTERASES

The nervous signal that causes a striped muscle to contract is mediated through the action of the substance acetylcholine, which acts upon a special receptor in muscle and which must be promptly removed after contraction to make it possible for the muscle to contract anew. Removal is brought about by the enzyme cholinesterase. An anticholinesterase is a substance that annuls the action of the enzyme that removes acetylcholine, the accumulation of which naturally leads to neuromuscular incapacity. The best-known anticholinesterases are the natural product physostigmine (eserine) and the related neostigmine. From the nature of their action, which includes potentiating the effects of acetylcholine at neuromuscular junctions, anticholinesterases bring some degree of symptomatic relief in myasthenia gravis and can to some extent annul the muscular relaxation produced by agents such as curare.

ANTIGENS AND ANTIBODIES

An antigen is a substance that arouses an immunologic response—typically of immunity, in which case the antigen is known as an *immunogen,* but sometimes of tolerance. The adjective "antigenic" is more reliable than the substantive "antigen," which is sometimes used to refer to the vehicle on which the antigen is carried. The word is also used rather loosely to refer to any substance that is recognized immunologically (that is, by its antigenic properties) even if the substance is not being used as or studied as an antigen.

When only part of a molecule is antigenic, that part is referred to as the *epitope* or *determinant group.* If the determinant group exists as a molecule in its own right, it is referred to as a *hapten,* a word often used with the tacit implication that a hapten is not antigenic unless it is attached to a larger molecule or is administered with an adjuvant that makes it so.

"Antigen" is thus defined operationally. The search for a substantive defining a property is complicated by the fact that chemical molecules of all kinds and of all shapes and sizes may be antigenic. For most purposes it suffices to say that the important defining characteristic of an antigen is that it is a nonself substance—or has become so by the kind of alteration referred to in the entry on IMMUNITY.

An *antibody,* on the other hand, is a modified blood protein aroused by the intrusion into the body of an antigenic substance and reacting specifically with the antigen or with that part of it which aroused the immune response. This reaction between antibody and determinant group is exquisitely specific; indeed, the resolving power of an antibody—its power to distinguish one molecular configuration from another—is unsurpassed among biological substances. So well attested is the specificity of antibody that distinctions which turn upon this property, as in the recognition of blood stains or of blood-group membership, are now accepted without demur as evidence in courts of law.

Antibody is the agent that puts humoral immunity into effect, especially antiviral immunity and some forms of allergy and sensitivity.

The ability to form antibodies is known to be essential for life because of the existence of a rare congenital abnormality, *agammaglobulinemia,* associated with a constitutional inability to manufacture blood proteins of the class to which antibodies belong. This disease was invariably fatal until the technique of electrophoresis was devised, which made it possible to recognize these proteins and therefore the pathological state in which they are lacking. Victims of agammaglobulinemia can survive only in the shelter of antibiotics.

APES

Linnaeus, the first great systematist of biology, grouped monkeys, apes, and anthropomorphs, including man, as the first among mammals and accordingly named them Primates. As Primates are first among mammals, so apes are first among Primates: the apes properly so called (the Barbary "ape," a macaque, is not) comprise orangutans (at one time given the same generic name, *Homo,* as man), gibbons, gorillas, and chimpanzees—these last two so alike in taxonomically usable characteristics that their allocation to different genera has been thought injudicious.

Our understandably human propensity to regard animal evolution as the impressive success story of which the denouement is man has made it natural to attach special weight to characteristics of apes that foreshadow the human condition—such as, obviously, the relatively large brain, the fully binocular vision, the beginnings of dexterity, the replacement of claws by nails and of an estrous cycle by a menstrual cycle related to periodic ovulation, and of course the loss of a tail. Social organization, too, of apes are clear adumbrations of the human condition: monogamy in gibbons, for example, and in many species the gathering of individuals into a band or troop that is often dominated by an older male. It cannot be said that the study of the sociology of apes has thrown a flood of light on the social structures of man. Rather it is the other way around: the social state of man provides valuable clues about what to look for among the apes.

The discovery of anthropoid apes in the fifteenth and sixteenth centuries may have produced a cultural shock less dramatic than it is represented to be in our discussion of MAN'S PLACE IN NATURE, because the notion of a great chain of being, deeply ingrained in Western thought, led more reflective people to expect a link (not, of course, an evolutionary link) between animal and man just as the so-called zoophytes were judged a connecting link between animals and plants. It is, needless to say, grossly unbiological to study apes as if they were prototypes or mock-ups of human beings. They deserve intensive study in their own right and are beginning to receive it.

The apes have not in a biological sense been enormously successful animals: the huge success of human beings that it would be perverse to deny is above all else the work of exogenetic evolution—that characteristically human form of social heredity which is mediated not through genetic but through cultural mechanisms.

AQUACULTURE

Aquaculture is to aquatic organisms what livestock farming is to terrestrial animals such as chickens and cattle. The animals water-farmed include fish such as catfish, trout, salmon, and mullet; crustaceans such as lobsters; mollusks such as oysters. A number of obvious parallels can be drawn between aquaculture and conventional farming. Both require the use of added fertilizers and nutrients that may be of organic or inorganic origin; both require the use of suitable housing or the provision of pens, very often in the form of cages or floating compartments bounded by nets or the suspended vertical lines with or without attached horizontal platforms used for oysters; both require strict control of predators and stringent procedures for the prevention of fouling. Aquaculture is, however, an intrinsically less manageable procedure than conventional farming if only because most—though not all—farmed aquatic animals are carnivorous and must be fed accordingly, in contrast to the relative simplicity of herbivorous animals that convert grain into animal flesh. Aquaculture should be thought of as a source of protein rather than of nutrients generally. The annual protein harvest is of the order of a million tons, with the prospect that improved techniques can increase this yield by half as much again.

ARCHAEOPTERYX

This creature is perhaps the best known of all missing links, forming a connection between reptiles and birds as convincing as that formed between reptiles and mammals by the duck-billed platypus and other monotremes.

Archaeopteryx, being feathered and probably warm-blooded, may be reckoned the first bird, dating from about 150 million years ago. Still, with teeth in both jaws, it is so like a small bipedal dinosaur that it would surely have been classified as such without comment if the imprint of feathers had not been recognized. It does not have a properly developed keel (the breast bone that in modern birds anchors the great pectoral flight muscles), so in all probability it was not much of a flier, being perhaps more of a glider.

The archaeopteryx was not the only reptile that took to the air: among others were the pterosaurs (including the famous pterodactyl), but the evolutionary modifications embodied in the genus *Archaeopteryx* were clearly a success. Not much more than fifty million years later, birds of modern characteristics, referable to existing orders, have come into existence.

ARISTOTLE

"Aristotle was 'a man of science' in the modern sense. He was a careful collector and observer of an enormous range of facts ... much of his work is still regarded with respect by scientists who

care to study it." These two sentences by the humanist Golds-
worthy Lowes Dickinson betray an almost majestic incomprehension of the character of science and of Aristotle's influence on science in the modern sense. A scientist is no more a collector and
classifier of facts than a historian is a man who compiles and classifies a chronology of the dates of great battles, major discoveries,
and so on.

Moreover, although Aristotle's philosophical opinions—on teleology, for instance—command respectful attention, the pioneers
of or spokesmen for the new science of the seventeenth century
(men such as Robert Boyle of *The Sceptical Chymist,* the Reverend
Dr. Joseph Glanvill, Francis Bacon the Lord Chancellor of
England, and the poet Abraham Cowley) repudiated the doctrinal authority of Aristotle with weary exasperation. No one did
more than Bacon to bring the doctrinal tyranny of Aristotle to an
end; Cowley wrote of the "pernicious opinion that all things to be
searcht in nature had already been found and discovered by the
ancients," and Henry Power, declaring that the new science was
"coming in on a spring tide," wrote, "Me-thinks I see how all the
old rubbish must be thrown away."

Aristotle's were the words that prompted Bacon to say that not
words but *works* were to carry the message of the new science; and
Aristotle's were doubtless the words that prompted the oldest and
most famous scientific society in the world to adopt as its motto,
from a couplet in Horace, *Nullius in verba*—Don't take anybody's
word for it.

Turning now to the biological works of Aristotle (which we
who do not have Greek must read in the fine translations published by the Clarendon Press in Oxford under the editorship of
J. A. Smith and W. D. Ross), we learn from the *Historia animalium*
that fish and birds severally are of many different kinds; that some
animals live in the sea and others on land; some are mobile, others
motionless; birds and bees have wings, other animals feet. Moreover, some creatures are tame and some are wild; some are at all
times tame, as man and the mule; others are at all times savage, as
the leopard and the wolf; and some creatures can be rapidly
tamed, as the elephant. Some animals make noises and others are
mute. There is nothing much to take exception to here, but elsewhere rumors and gossip take over: some birds are peculiarly "salacious," as the partridge, the barn-door cock, and their con-

genors; others are inclined to chastity, as the whole tribe of crows, for birds of this kind "indulge but rarely in sexual intercourse." Some animals are good tempered, others ferocious, and still others mean and treacherous; but the lion is noble and highly bred.

The biological works of Aristotle are a strange and generally speaking rather tiresome farrago of hearsay, imperfect observation, wishful thinking, and credulity amounting to downright gullibility. What empirical evidence can have convinced Aristotle that the semen of youths between puberty and the age of twenty-one is "devoid of fecundity" and that young men and women produce undersized and imperfect progeny? Prolapse of the uterus and menstruation as often as three times within a month are said to be symptoms of "excessive desire."

It is hard not to sympathize with those new scientists of the seventeenth century who resented and repudiated the doctrinal authority of such statements. We can get some insight into the origin of these bizarre opinions from Sir Philip Sidney's *Apologie for Poetry* (1598). Sidney deplored the bondage of history to the empirical and the particular, and declared that the historian's "bare *was*," as Sidney dismissively described it, could never have the force of poetic truth; for poetic truth in Aristotle's conception was a revelation of the ideal, of what *ought to be*. For Aristotle poetry was a more philosophical and a higher thing than history. Poetry could reveal what ought to be in the light of an understanding of Nature's true intentions.

There can be little doubt that Aristotle's conception of history extended also to science—and in this light we can understand some of his more strangely erroneous opinions. Aristotle was a firm believer, for example, in the Hebdomadal rule, that everything goes in sevens. Man has seven ages, each seven years long, so anyone with a true understanding of nature would realize that human semen must be infertile between the ages of fourteen and twenty-one, for would not fertility in this period contravene the Hebdomadal rule? And so what ought to be became what was. Sometimes of course Aristotle is right; his writings were so voluminous he could hardly fail to be correct sometimes (irreverent thoughts of monkeys and typewriters steal into the mind). Thus Aristotle roundly declares that the semen of the Aethiops is not black and chides Herodotus for thinking that it is.

We do not believe that anyone who decides not to read the works of Aristotle the biologist will risk spiritual impoverishment. For nowhere in Aristotle do we find the inspired insights of a Samuel Taylor Coleridge, whose *Hints Towards the Formation of a More Comprehensive Theory of Life* (London, 1848) can still be read with enjoyment and profit by anyone familiar with the state of the biological sciences when he wrote.

ARTHROPODS

The most various and the most numerous of all animals, arthropods are by these criteria the success story of the animal kingdom. The major classes are the insects and the crustaceans, which between them have adapted to all environments. The jointed limbs, other appendages, and indeed body, that give the phylum its name are necessary by virtue of a tough exoskeleton of which the principal constituent is *chitin,* a flexible but virtually indigestible substance distantly related to cellulose. Chitin forms the outer casing of the entire body. The growth of an arthropod is made possible only by periodic molting (technically, "ecdysis"), during which the exoskeleton is shed and replaced by a newly formed skeleton.

Crustaceans and insects, unlike vertebrates, do not have a so-called closed blood vascular system. The blood, instead of being contained within blood vessels and propelled by the action of a pump (the heart), percolates freely through the body cavity and the interstices of tissues. It is propelled at very low hydrostatic pressures by the peristaltic contractions of a tube charitably called a "heart," which lies dorsally (in contrast to the ventral heart of vertebrates). An open blood system such as this hydrostatically requires a fairly resistant outer skeleton if there is to be any circulation of hemolymph. The situation is much the same as in vertebrate animals, where the arteries—especially those near the heart—must be strengthened by a layer of circular muscle.

An exoskeleton confronts insects with a special respiratory problem, overcome by the direct piping of atmospheric oxygen to the tissues through a branching system of exquisitely fine tubules called *tracheae.* This mechanism is so slow and inefficient that we suspect that bodily movement such as the beating of the wings contributes something toward respiration.

Another morphological characteristic that distinguishes the arthropods and the notable line of descent of which they are the principal members is that their nerve cord is not dorsal as it is in chordates, but runs down the ventral midline. (This explains why students are always taught to dissect invertebrates from the dorsal side.) The segmentation of an arthropod is not, as it is in chordates, primarily a segmentation of the longitudinal musculature; it is a whole-body segmentation, each segment containing a nerve ganglion, an excretory organ, and usually appendages.

Arthropods must have descended in evolution from an annelidlike ancestor. Members of the arthropodan class Onychophora, of which peripatus and peripatopsis are representative members, are intermediate in characteristics between arthropods and annelids. Peripatus is thus a so-called missing link, or, for more literary minds, a rung in the great scale of being.

ATAVISM

When John Langdon Down in 1866 described as mongolism the developmental disorder now known to humane and discriminating people as Down's syndrome, he did so because he regarded the condition as degenerative—as the consequence of reversion to a remotely ancestral human condition, supposedly similar to that of modern Mongolians, or (as he implied in a way that did not endear him to Mongolian members of the World Health Organization) to an "even lower" condition. A victim of Down's syndrome was thus seen as a throwback, as an example of the phenomenon of atavism; likewise the Italian criminologist Cesare Lombroso (1835–1909) interpreted the "criminal mind" as a consequence of atavism, a throwback to a lowly and primitive human condition. This was an idea that fitted well with the dark and confused thoughts on heredity entertained by literary people in Victorian and Edwardian times, and by "thinkers"—if that is not too farfetched a description—such as Max Nordau, whose popular and influential work *Degeneration* (1898) was dedicated to Lombroso. Nordau reported that Richard Wagner, Henrik Ibsen, and Oscar Wilde all were to various degrees degenerate.

The notion of atavism clearly grew out of a popular misconception of the law of recapitulation, a principle no longer accepted in the form in which Ernst Haeckel propounded it.

From a genetic point of view the reappearance of an ancestral trait, especially a recessive trait in remote descendants, is by no means impossible; Aristotle recognized this as a possibility and gave examples. A chance mutation or merely the luck of mating may bring together two recessive genes that have remained apart for many generations; it is an important theorem of Mendelism, confirmed by all subsequent research, that genes are physically and genetically stable entities which may be relied upon to exercise the same effects in the same genetic and developmental environment. Thus it is to Mendelism that we must turn for an explanation, and Mendelism that alone provides it. Throwbacks of the kind described must often have been remarked by stockbreeders. The phenomenon as it applies to single genes or gene pairs is not wildly improbable; on the other hand, the likelihood that an entire ancestral genome will be reconstituted in a grandfilial or later generation is of the same order as the likelihood that a kettle placed on ice will boil instead of cool (the example chosen by Arthur Edington to illustrate the probabilistic character of the Second Law of Thermodynamics).

BACTERIOPHAGES

A bacteriophage is a virus that infects bacterial cells, often causing their dissolution (lysis). This property at one time raised high hopes for the therapeutic usefulness of bacteriophages as antibacterial agents, a possibility that played an important part in Sinclair Lewis' 1925 novel, *Arrowsmith*. Best known are the bacteriophages of the T series, which infect the organism *Escherichia coli* (*Bacterium coli*).

The study of bacteriophages has made an invaluable contribution to molecular genetics; information was secured in months that might have taken several decades had experimentalists been confined to the use of viruses that infect ordinary tissue cells in tissue culture. Thus research on bacteriophages has been in the highest degree heuristic—a word which laymen more often look up and more often forget than any other in the philosophical vocabulary. If Webster's is anything to go by, the term means "serving to guide, discover, or reveal."

BARNACLES

Barnacles are ubiquitous marine animals familiar to anyone who has visited the seaside or inspected the hull of a seagoing vessel; they are a subclass of crustaceans and are thus related to crabs, lobsters, shrimps, and the like. Anatomically speaking, they are attached at the head end, and the shell—a carapace—is strengthened in such a way as to make the barnacle virtually impregnable. Respiration and feeding are carried out by a continuous clawing motion of the rather feathery jointed legs that give the group its zoological name: Cirripedia (feathery feet).

Barnacles are found not only on rocks and submerged logs, but also on the shells of mollusks and the larger crustaceans; they can attach themselves to the teeth of whales that have teeth and to the whalebone filters (baleen) of whales that do not. Aristotle understood the special reproductive problems of creatures which do not move and very understandably formed the opinion that barnacles "have no sex any more than plants have, but as applied to them the word is used only in virtue of a similarity and analogy." Most barnacles are hermaphrodite, though cross-fertilization is usual; in some barnacles, indeed, the male sex is reduced to a mere reproductive organ that is parasitic upon the female.

Barnacles are to outward appearances so different from crabs and their relatives that one may wonder at the complete confidence with which they have for many years been classified as crustaceans. The decisive clue is their possession of a larva of a kind characteristic of crustaceans—the nauplius, which serves the purpose of distribution (another understandable problem of sessile animals).

Not all barnacles are sessile, however, and one particularly graceful "goose" barnacle with a white shell about the size of a small mussel carried on a curving stalk was thought in the Middle Ages to give birth to geese. While it is highly unlikely that Linnaeus thought so, he gave the notion countenance by allotting them Latin names that embodied this conception: *Lepas anatifera* or *L. anserifera,* duck-bearing and goose-bearing, respectively. Charles Darwin wrote a massive treatise on barnacles between 1851 and 1854, but later came to have doubts (which we share) about the wisdom of having embarked upon such a tedious and unilluminating enterprise.

Some barnacles are parasitic and have undergone such radical modification in the course of becoming so that they are barely recognizable as barnacles, let alone as crustacea; these, too, are given away by their larvae, which are characteristically crustacean.

Because of the degree to which they foul seagoing vessels and almost anything else that is long immersed in seawater, barnacles are a relatively severe economic liability and accordingly are assaulted by ultrasound and a variety of poisons; environmentalists need feel no concern, however: the barnacles are sure to win.

BEHAVIORISM

Regular novel readers must be aware that novelists are of two kinds: on the one hand, there are novelists who, in addition to writing a narrative of events, arrogate to themselves the privilege of looking into their characters' minds and reporting what they find. Such novelists tell us what their characters are thinking and what their feelings and emotions and hopes and fears are. Thus we can connect the narrative with the characters' intentions and interpret their interactions in terms of this privileged insight into what they feel about themselves and about one another.

The other kind of novelist tells us about his or her characters only what we ourselves should observe or infer about them if somehow we were ubiquitous spectators of everything that was going on. In addition, then, to narrative—that is, "story"—these novelists describe how their characters act and what they say to one another in a variety of special circumstances, generally without using the subterfuge of recording what they are saying to themselves. No special attention is paid to the characters' thoughts and feelings: these only become apparent through what they do or say.

Novelists of the former kind can be likened to the "introspective psychologists" who at one time dominated psychology, whereas novelists of the second kind (among whom Jane Austen is the supreme example) can be more accurately likened to behaviorists. Given the pen of a technically brilliant writer who is also a shrewd and attentive observer, the behavioristic style is enormously effective. We do not need Jane Austen to spell out for us that Mr. and Mrs. John Dashwood are utterly mean and self-centered; she bares their small souls in the first half-dozen pages of *Sense and Sensibility* in a way that no psychologizing could improve upon.

Behaviorism, now three-quarters of a century old, is at once a philosophy of mind and a methodology of research. Let us consider first its methodological aspects. John Broadus Watson (1878–1958) and those persuaded by him carried out a virtually Baconian revolution in psychology, at all points substituting the empirical for that which, because it was not presented to the senses, could only be known by inference—especially states of mind such as joy, misery, or malevolence, or indeed (for where do we draw the line?) consciousness itself. This methodology is profoundly influential and has supplanted the presumption of privilege in introspective psychology by empirical narrative and reportage.

When we look at it as a philosophy of mind, we see that behaviorists have been vastly misrepresented—above all as advocates of a rudely mechanistic reduction of man to the status of a skillful performing machine without mind, without psyche, and without that special will to communicate, or outgoingness, which to ordinary people is evidence of sentience as convincing as body warmth and heartbeat are evidence of physical life.

In imputing these dark thoughts to behaviorism, its critics are confusing the methodology with the philosophy of mind. A behaviorist would not deny that consciousness exists; he would not carry much conviction if he did (such an utterance would itself convince his audience of his own consciousness). However, he would positively affirm that there is no *need* to use the concept of consciousness in psychological analysis. Instead of using concepts that are inferences drawn from our empirical awareness of a mode of behavior, he would ask why we do not content ourselves with merely recording behavior. This is a Jane Austenish way of going about things.

For these misunderstandings of behaviorism and the general air of odium which tends to surround them, behaviorists have only themselves to blame. Much behaviorist writing is marked by an aggrieved and truculent air, as if a principal objective of the new science were to put down once and for all the presumption of privilege and the unduly literary pretensions of introspective psychology.

That particular battle has been won: there is no need for behaviorists to continue it. The philosophy of mind element in behaviorism found an unexpected recruit in one of England's foremost philosophers, the late Gilbert Ryle of Oxford University, whose 1949 *Concept of Mind* had (he did not deny) a strongly behaviorist coloration.

Despite its contribution to methodology, behaviorism cannot be said to have produced any significant conceptual revelation or, considered as a psychology, to have either deepened our understanding of the mind or enlarged our armory of preventive or curative procedures.

BILHARZIASIS

This debilitating disease (now more often called schistosomiasis) is caused by the invasion of the blood vessels of the viscera and even of the brain by "flukes," one of more than two thousand species of parasitic flatworm. The flukes in bilharziasis were first recorded by Theodore Bilharz, a German surgeon working in Cairo, after whom the disease was originally named. Worldwide, its victims number in the millions: the disease is recorded in Africa, Japan, South America, and the Caribbean. Its damaging effects are partly mechanical in origin and partly the consequence of the intense allergic reaction the parasite excites. The intermediate host is a water snail which is not itself grievously afflicted, so it will not do to describe schistosomiasis as a disease of water snails propagated by means of a human vector. The control of spread is to a large extent behavioral: members of a population at risk therefore are urged not to bathe in snail-infested waters.

BIOENGINEERING

No one description of this scientific or paramedical area of inquiry would satisfy all its practitioners. Bioengineering comprehends several very different activities.

First among them is the devising of machinery that can perform or temporarily deputize for physicomechanical functions of the body. One example is the artificial kidney which performs the ultrafiltration that is a principal part of the kidney's function; a second is the heart-lung machine that has made possible open-heart surgery and heart transplantations. Various artificial limbs, even hearing aids, qualify.

Another aspect of bioengineering is the analysis of physicomechanical aspects of the working of the body. Movements such as swimming, walking, and running are scrutinized, and thermodynamic balance sheets having to do with energy intake and expenditure of energy through muscular movement and loss of heat are drawn up.

Bioengineering also describes the use of biological methods in manufactures and in industry generally. Wine making, the dressing of leather, the preparation of enzyme-containing detergents, and the making of vinegar are all biological enterprises, many of them discovered and worked out empirically.

It is not through these straightforward and beneficent forms of bioengineering that the Wicked Scientist proposes to put into effect his scheme for abolishing consciousness and the psyche, and for diminishing man to a mindless but obedient automaton whose interactions with his fellowmen, no longer obfuscated by love or feeling, will be confined to the exchange of factual information. No—it is by means of *genetic engineering* that the Wicked Scientist will try to achieve his ambitions. Genetic engineering is all that goes into the deliberate alteration of the genome in a predetermined direction.

It may come as a shock to the layman to learn that we already have in our power the means to bring into existence two different kinds, or *marques,* of human beings: one tall, handsome, and well endowed physically and intellectually; the other stunted, dumb, and obedient, with the appearance, submissiveness, and capabilities of a serf. Such human beings would differ from each other no more than some breeds of dog already do. All that is needed to bring about this end result is time, money, and inclination. Human beings have an open, "wild-type" breeding system with a vast store of genetic variance. Nothing in biological principle stands in the way of our applying the most powerful known technique of genetic engineering, artificial selection, to treat people, in effect, like dogs; for this is how we *have* treated dogs. (See P. B. Medawar, *Pluto's Republic* [Oxford, 1982], p. 311.) To do so would require tyrannical enforcement of a policy of selective breeding over a period no longer than the duration of the House of Hapsburg. All that has stood in the way of achieving this ambition has been the worldwide shortage of mad scientists and the virtual impossibility of political policy's remaining consistent through a large number of human generations.

"Genetic engineering" normally connotes something rather different: it implies deliberate genetic changes brought about by the manipulation of deoxyribonucleic acid (DNA), the vector of hereditary information. Into the DNA of a suitable host (a nonvirulent bacterium, for example) must be inserted a length of DNA embodying the genetic instructions that code for synthesis of the desired end product. This might be a hormone such as insulin or growth hormone, or an antigen that we use for vaccination. The feat has been made possible by a whole series of superbly accomplished technical discoveries that provide the experimenter with an arsenal of machine tools permitting interruption of the DNA chain and splicing into it of the requisite length of DNA.

Some recent developments are decidedly encouraging, though the millennium is not yet as near as some advertisers would like us to believe. People old enough to read this book will surely remember the fanfares that heralded the Age of Plastics—fanfares that continued till people got fed up with being told how in the imminent future virtually everything in the world would be replaced by a plastic substitute. It was wearisome indeed; but plastics came just the same, and mighty useful they have been. Vicariously manufactured hormones, antigens, and pharmacologically active agents will come along too. It would be reckless to deny it, for is it not a major truth of technological history that anything which is in principle possible will be done if the intention to do it is sufficiently resolute and long-lasting? Land on the moon? Yes, assuredly. Abolish smallpox? A pleasure. Make up for deficiencies in the human genome? Mmmm, yes, though that's more difficult and will take longer. We aren't there yet, but we're certainly moving in that direction.

Some science writers, irresponsible members of an otherwise scrupulous and rather important profession, have represented these impressive experimental exploits as steps preparatory to a sort of biological Grand Guignol in which all manner of genetic and epigenetic enormities will be perpetrated. Nothing in the past history of biology justifies the calumny that biologists have equipped themselves to perform and exult in the possibility of performing such tricks simply because modern bioengineering has made them possible. One of England's most famous geneticists, upon his return to England, was quizzed by a customs officer, then told, "I suppose I'd better let you through, otherwise you'll turn me into a frog." At least the agent demonstrated a proper sense of his station in life, as if instinctively aware that anyone so credulous and gullible did not deserve to be turned into a fairy prince.

BIOGENESIS

In its affirmative form, the Law of Biogenesis states that all living organisms are the progeny of living organisms that went before them. The familiar Latin tag is *Omne vivum ex vivo*—All that is alive came from something living; in other words, every organism has an unbroken genealogical pedigree extending back to the first living things. In its negative form, the law can be taken to deny the occurrence (or even the possibility) of spontaneous generation. Moreover, the progeny of mice are mice and of men, men—"homogenesis," or like begetting like.

The Law of Biogenesis is arguably the most fundamental in biology, for evolution may be construed as a form of biogenesis that provides for the occasional begetting of a variant form.

Not only whole organisms but some of their parts such as the mitochondria are biogenetic too—that is, they are not synthesized de novo in the course of development but are formed from the existing mitochondria, of necessity transferred in sperm or egg.

BIOLOGY IN MEDICAL EDUCATION

Because so much of medicine is applied biology, it might be judged a self-evident truth that biology has exercised a penetrating, wide-ranging, and wholly beneficent effect on medical education and on the physician's thinking. In reality this has not been so. Biology in an old-fashioned sense (that which would have been recognized by Charles Darwin) has imposed an almost intolerable pedagogic burden on medical education and has been deemed responsible, too, for such grave fallacies as the Panglossism we shall discuss below.

It is indeed a profoundly important lesson that a human being is an animal, but it was not a biologist who taught it. Aristotle was aware of it, and it is implicit in Plato's great chain of being.

In any event, it was a London physician, Edward Tyson, who in 1669 demonstrated for all to see the affinity between monkey, man, and a baby chimpanzee (a "pygmie").

In the half-century following the publication of Darwin's *Origin of Species*, the great evolutionary biologists of the nineteenth century—men such as Karl Gegenbaur, Johan van Wijhe, and Edwin Goodrich—saw it as the principal task of biology to validate the evolutionary conception and to demonstrate its unfolding, mainly by the methods of comparative anatomy. Fired by evangelistic zeal, they were not satisfied with working out the grand dynasties of evolution, but spent much of their time searching in the parish registers of evolution and attempting to trace descent among orders and suborders.

It soon came to be thought that no one could truly comprehend man without a clear understanding of the evolutionary process of which man was the supreme product—unless, indeed, the doctor-to-be was taken on a conducted tour through the animal kingdom, starting with *Amoeba* and pausing en route at earthworm, cockroach, dogfish, bony fish, frog, and rabbit. By the time he had finished, he supposedly would have a deep historical understanding of what would confront him when he went into the dissecting rooms of the anatomy department. So began the weary grind of what became in the United Kingdom the propaedeutic year known as "first MB," which postponed for yet another year a medical course whose consummation took far too long anyway. Things were never so bad in the United States, because the liberal arts tradition could make sure that the biology a future medico learned was part of a well-designed course that was educationally valuable in its own right, irrespective of whether the student who took it went on to medicine, science, or any other professional pursuit.

It is not high praise, but it deserves to be said that in all probability few medical students' minds were ever irreparably damaged by taking the first MB course or something serving the same purpose. Perhaps the principal mischief of these propadeutic exercises was to create the illusion that biological observations of plants and animals have a special authenticity that has led to the depreciation of human biology. Today, more enlightened, we realize that theoretical genetics is as cogently taught upon man as upon fruitfly or pea or mouse. Indeed, in many ways man is to be preferred, because the genetics of man lends itself particularly well to the realistic study of polymorphism, population genetics, and the like.

Another mischief that must be debited to propadeutic courses in biology is their arousal in many medical students of a lifelong immunity to biology that closes their minds to its riches and intellectual delights. Sometimes, too, ill-understood lessons on Darwinism and on evolution theory generally have given rise to that peculiarly medicobiological aberration of thought that has come to be called Panglossism, after the egregious Dr. Pangloss of *Candide*. The message Voltaire pulled the carpet from under was *Tout est bien:* All is well; Nature knows best, and natural dispositions are necessarily for the best. Have they not, after all, been tried in the fires of natural selection, so that we are left with the best possible, if not always the best imaginable, solution of the manifold problem of remaining alive and reproducing in a pretty thoroughly hostile environment?

Consider, for example, the gross ineptitude of the healing of injuries that lead to extensive loss of the full thickness of the skin—those that might result from a severe mechanical injury or from a deep and extensive burn. The "natural" process of repair starts, as a rule, with the sloughing of dead tissue and its replacement by temporary tissue of repair known as granulation tissue, a highly vascular proliferation of connective-tissue cells and blood vessels that fills the area from which tissue has been lost. Repair proceeds by the creeping of normal epidermal cells over the raw granulating surface, starting from the edge of the wound and from any islands of living epidermis left within. This newly formed epithelium is mechanically weak and extremely unstable. The granulation tissue underlying this temporary surface turns into fibrous tissue by the deposition of connective-tissue fibers, but a true skin of three-dimensionally packed connective-tissue fibers never regenerates; nor, as a rule, do sweat glands, oil glands, or hairs. Worse still, the wound now undergoes a process of contraction that approximates the original edges, often leaving a scar that outlines the wound edge. This process of contraction is something greatly to be feared: it is disfiguring and otherwise disabling. On the leg, for example, the contraction may easily interfere with the arterial blood supply. Indeed, contraction almost always impedes lymphatic drainage and venous return. The end result is functionally inept and as a rule cosmetically hideous.

The remedy is skin grafting, a vast improvement upon Nature, by means of which the loss of body fluids from the raw area is arrested, the danger of infection is reduced to a minimum, and above all contraction is forestalled and with luck largely circumvented. It is not to the reconstructive surgeon that one should preach the text that Nature knows best.

BLOOD AND ITS CIRCULATION

A salty, viscous, yellowish or straw-colored fluid, blood contains a variety of corpuscles in suspension, especially the red blood cells (*erythrocytes*) whose alternate oxygenation and deoxygenation is responsible for the transport of oxygen throughout the body. The fluid part of the blood is known as blood *plasma;* but when blood clots, the fibrin fibers of which the clot is formed contract and express a nonclotting fluid known as *serum,* which is normally free of contained cells.

Although there are exceptions, arteries usually contain oxygenated blood under a pressure that is estimated by ascertaining with a simple apparatus the applied pressure necessary to occlude an artery and temporarily stop the blood flow. Veins in turn contain deoxygenated blood, which is darker than arterial blood—the exception being, of course, the pulmonary artery, which contains not-yet-oxygenated blood to be conducted to the lungs, from which the pulmonary veins will return oxygenated blood to the atrium of the heart. Arteries generally have muscular coats, and it is the contraction of these coats that prevents bleeding when small arteries are cut.

Apart from red blood corpuscles, blood plasma holds in suspension a variety of white blood corpuscles, or leucocytes, the most numerous among which are the lymphocytes and the *granulocytes* of various kinds, which can ingest foreign particulate matter such as bacteria (especially if their targets have been coated beforehand with antibody).

* * *

It is difficult nowadays to put oneself in the position of those who at one time believed that the movement of blood is essentially an ebb and flow—an advance and retreat. Thus it is easy to underestimate the importance of the discovery of the circulation of the blood, a discovery which by its sheer magnitude and by the style in which it was made earned William Harvey a place on the team of All-Time Greats captained by Sir Isaac Newton, which counts among its members Galileo, Darwin, and those few others who in their time changed the direction of the flow of thought.

Although the Spanish physician Servetus recognized the pulmonary circulation (from heart to lungs and back again), Harvey went far beyond him in discerning the nature of the pulse and the anatomic basis of the systemic circulation: he perceived, too, the function of the valves, the passive nature of the dilatation of the heart—and indeed recognized all the fundamentals of our present understanding of the circulation of the blood.

Harvey discovered all this by anatomic observation and by simple and sufficient observations—the best kind.

It is most unfortunate that in order to give countenance to a favorite mystique having to do with clinical observation, and to make Harvey appear "scientific" in the style that John Stuart Mill and Karl Pearson believed to be distinctive of science, Harvey has been portrayed as an inductive scientist who, with no preconceptions whatsoever, made observations and conducted experiments from which he deduced the theory of the circulation of the blood. We believe that in reality Harvey must have had a clear preconception of the notion of circulation. So, far from collecting facts and piling up empirical observations, his experiments were exactly those that would have been carried out by someone testing an existing hypothesis. In short, Harvey was much more of a card-carrying scientist than he is represented to be by those who have tried to claim he was an inductivist who did no more than attend humbly to Nature's lessons as he learned them from her own lips.

The evolutionary history of blood has been the subject of a number of romantic speculations, one of which was that blood is essentially an evolutionary descendant of seawater and that its salty composition in mammals today reflects the salty composition of the sea in the era during which fish left the sea to colonize dry land. There would be some substance to this notion if there was even the faintest likelihood that the blood of fish was ever in osmotic and ionic equilibrium with the seawater that surrounded them. There is a certain affinity between the salty makeup of seawater and that of blood; inasmuch as life must have begun in the sea, the metabolism of cells must surely have evolved in relation to the inorganic constituents of the sea, especially sodium, potassium, calcium, and magnesium cations and chloride anions.

The hypothesis of a marine origin of blood has been cited as an example of *poetism,* that is, the evaluation of scientific hypotheses by literary rather than scientific criteria—a miscarriage of thought as offensive as *scientism,* the application of supposedly scientific procedures to the investigation of matters on which science has no bearing whatsoever (P. B. Medawar, *Pluto's Republic* [Oxford, 1982], pp. 60–61).

The term "blood relationship" is a figure of speech for kinship, for genetic relationship. Although there is no foundation in theory or fact for the belief that the characteristics of blood give a deeper insight into affinities than the characteristics of other tissues, because of the accessibility and forensic significance of blood, its properties are the ones studied most intently in problems having to do with identity. The methods of ascertainment are less often chemical than immunologic, because immunologic methods have a higher resolving power and a higher degree of specificity than any other.

It goes without saying that blue blood is not a sign of gentility, nor red blood of fortitude and courage. Nevertheless, in some contexts the figurative meaning has wholly usurped the literal. "Sanguine" is never used today to signify blood-red, but only the cheerfully hopeful disposition to which a ruddy countenance is thought to bear witness. The word "optimistic" is fastidiously shunned by those reluctant to impute to the person to whom it refers a belief in some of the metaphysical consequences of the philosophy of Leibniz.

CANCER

The branch of pathology that has to do with the study of cancer is called *oncology* (the prefix *onco-* always relates to cancer, as in oncogen or oncovirus). What we know about cancer is the accumulated understanding of many centuries of clinical observation and about a hundred years of systematic research.

Some pathologists, oppressed by the enormous variety in cancers, their modes of growth, and the ways in which they may be caused, deny that cancer is a clinical entity: it is not one, they say, but many diseases of different origins and different behaviors, for which we should expect to devise a proportionate number of different treatments. While not denying any of this, we do not think we are yet knowledgeable enough to refute the more straightforward assumption that there is a commonality in the formation of all tumors—and that therefore there is a fundamental resemblance among them all, though it has not yet been discovered. This is not to imply that there need be any such thing as "the" cure for cancer. Individual tumors are sure to have distinctive properties, such as dependence upon specific hormones, of which it is often possible to take clinical advantage.

In the accepted terminology of cancer the suffix *-oma* usually refers to a nonmalignant growth or swelling: a corn, for example, produced on the dorsum of a toe by tight-fitting shoes is an epithelioma, and the benign swelling of bone is an osteoma. When a growth of epithelium is malignant, it is referred to as a carcinoma; some purists insist that the word "cancer" should not be used as a generic term for all malignant tumors but should be confined to tumors of epithelial origin. Usage, however, is tending to annul these sematological niceties. The word "sarcoma," for instance, is used for a malignant growth of skeletal, connective, or circulatory tissues, as in osteosarcoma, fibrosarcoma, lymphosarcoma.

A rapid growth rate is not a universal characteristic of tumors, nor is it the property most to be feared: it is the *invasiveness* of tumors that makes them simultaneously more dangerous and less accessible to surgical treatment. *Metastasis* or the spread of the tumor throughout the body from its place of origin is another serious complication.

Even though cardiovascular disease long ago became the principal cause of death of adults in the industrialized Western world, cancer still arouses more anxiety than any other disease. Lewis Thomas has called attention to the parallel between the dread of cancer and the dread of tuberculosis ("consumption") that was prevalent at the turn of the century and for a decade or more afterward. The diagnosis of consumption too was construed as a death sentence, especially if the infection had spread throughout the body.

This attitude toward cancer is widely prevalent today: cancer, it is thought, is incurable, progressive, inevitably fatal, and never remits of its own accord. These beliefs are not entirely accurate: cancers *can* sometimes be cured, although admittedly not as often as is implied by articles in magazines whose editorial policy is doggedly optimistic in order not to alienate readers and advertisers alike. The combination of surgery with chemotherapeutic drugs does arrest the growth of some cancers sometimes; still, there is nothing to justify a sanguine presumption of success in all cases.

Moreover, some cancers do undergo spontaneous regression, do dwindle away and disappear even without therapeutic intervention. This phenomenon, combined with the belief that many more tumors arise than ever become a clinical threat, has given rise to the widespread opinion that there is some natural bodily defense against cancer. Many biologists believe this, although the biological case for such a defense system's having evolved is rather weak. Unlike infectious disease, cancer has not been a cause of mortality great enough to have generated much selection pressure: it is largely a disease of postreproductive life—that is, of a period during which (as explained in the entry AGING) the force of natural selection is greatly attenuated.

What form could a natural defensive mechanism take? The consensus is that if such a mechanism exists it is in a general sense immunologic and depends upon the tumor's being endowed with or acquiring sufficient nonself properties to be antigenic. In laboratory animals tumors caused by oncogenic viruses and oncogenic chemicals are almost invariably antigenic and often arouse an immunity reaction strong enough to cause them to be rejected, that is, to dwindle away. It is less certain that this is true of the tumors described as "spontaneous" because their causes are unknown.

Applied cancer research is mostly a matter of seeking new and better antiproliferative drugs. The difficulty in principle is that none of these drugs is cancer-specific. They retard the growth of proliferating cells but may themselves be oncogenic—as X rays and gamma rays are well known to be. Because of these drawbacks, the notion of an immunologic treatment is especially attractive but the results so far have been rather disappointing.

One of the most interesting new areas of cancer research has grown out of the notion of *anaplasia*. As this notion is of metaphysical origin, it may constitute an especially apt example of Karl Popper's thesis that far from being dismissed as so much nonsense, metaphysical speculation as a source of ideas is entirely admissible in science provided those ideas are exposed to the rigorous critical examination to which all scientific notions should be exposed.

In the nineteenth century in Germany nature-philosophical speculation affected clinical pathologists as well as conventional biologists. Some pathologists thought there was a deep-seated connection between cancer cells and embryo cells. With the neurone in mind, they reasoned that a cell's ability to multiply varied inversely with its state of differentiation: the more highly differentiated the cell, the less able it was to undergo division (cells of the early embryo obviously can and highly differentiated cells such as neurones obviously cannot). Because cancer cells are readily able to divide, they must, it was argued, have undergone dedifferentiation and reversion to an embryonic condition. Cancers, Julius Cohnheim thought, developed from embryonic "rests"—or leftovers—from embryonic cells which because of some misadventure had failed to differentiate and had persisted anomalously into adult life. This is a hypothesis long since experimentally falsified, though one can understand its nature-philosophical appeal. Moreover, dedifferentiation does not occur:

most cancer cells retain their differentiated properties and continue to manufacture their characteristic secretory products. Anaplasia, however, does occur in a somewhat modified sense: cancer cells may manufacture anew substances produced in embryonic or fetal life by their own distant cellular ancestors. It is to this unexpected reawakening and reactivation of genes which functioned earlier in embryonic life that the word "anaplasia" is now applied.

In essence cancer research is based upon an attempt to find a clinically usable difference between a malignant cell and its normal counterpart. Anaplasia is not the definitive answer to this problem, but it is the closest approach so far. The phenomenon *has* aroused special interest, and a recent conference on fetal antigens in cancer convened by a charitable foundation in London was attended by cancer research workers from nine different countries. Cancer research is by no means barren of ideas, as some seem to suggest when seeking excuses to reduce its handsome subsidies or diminish its standing in science. We believe that cancer will eventually be brought under control by the further development and medical application of ideas current today.

No one will deny the truism that the best way to cope with the cancer problem is not to get cancer. It should be a source of some rejoicing, therefore, that current expert opinion puts as high as 80 percent the proportion of cancers that arise from exogenous, and therefore in principle preventable, causes. We have in mind cancer viruses, industrial chemicals, ionizing radiations, asbestos fibers, and many food additives (preservatives, "improvers," coloring matter, and flavoring). Although it is not always possible to equate the sanctity of human life with the sanctity of profitable commercial enterprise, vigorous legislation is being enacted in both the United States and the United Kingdom to insist upon safeguards with respect to commercial use of potentially oncogenic chemicals.

For the past quarter-century the most valuable evidence relating to the causation and control of cancer has come not from the clinic or the laboratory but from epidemiologists in league with demographers, the registrars of births and deaths. The culture hero of cancer epidemiology is the Percival Pott who first demonstrated the connection between cancer of the scrotum and the occupation of chimney sweep. Reasoning similar in principle has subsequently established a connection between oncogenesis and radioactivity, between skin cancers and coal tar and its distillation products, between smoking and lung cancer—and more recently between cancer and deficiency of vitamin A and the carotenoids.

The order of magnitude of the sums spent to promote cancer research is hundreds of millions of dollars, but the value of that research is not to be estimated solely with reference to its effects on the management of cancer: the fallout is considerable. For example, the transplantation of tissues and organs was made possible by the judicious use of immunosuppressive drugs secured by raids on the storehouse of chemotherapeutic agents devised for use in the treatment of cancer (among them azathioprine, Imuran, and methotrexate).

Epidemiology has also thrown important light on the relationship between sex, reproductive history, and cancer. For almost all cancers married women are less at risk than single women—and women generally have lower cancer rates than men. In women, moreover, there are clear correlations between cancer risk and reproductive history. Cancer of female reproductive organs is more common in childless women than in women who have had children and its frequency falls as parity (number of children born) increases. In both North America and the countries of northern Europe, ovarian cancer reached a peak in the generation of women born in the quinquennium 1895–1900, in which fertility was so low that demographers of the thirties wondered whether people of the Western world might not die out. There is also a most interesting correlation between age of a mother at birth of her first child and risk of contracting breast cancer: teenage mothers have less risk than mothers who were twenty-five or more when their first child was born; and women primiparous at thirty or older have a decidedly higher risk of breast cancer than childless women.

CAROTID ARTERY

This is the major blood vessel in the neck that must surely be intended when inexpert thriller-writers refer to the "jugular." The internal and external carotids are the principal vessels transporting oxygenated blood from the heart to the head; it is occlusion of the carotid that causes rapid loss of consciousness, and the carotid whose inadvertent puncture by a surgeon is sure to be the subject of unfavorable comment.

Thus when Bulldog Drummond—hero of popular adventure stories and much admired by schoolboys and retarded adults—is said to have grasped and pressed the jugular vein, judicious editing will at once substitute "carotid artery."

CELL THEORY

The smallest subdivision of a tissue that is capable of independent life when kept alive in tissue culture is a cell. It may be thought of in alternative ways—on the one hand as the unit from which a tissue is compounded, and on the other hand as a mechanical or, one might say, administrative subdivision of the tissue. "Administrative" because from the standpoint of genetics and epigenesis a cell is the sphere of influence of its nucleus. Although some cells (mammalian red blood corpuscles, for example) have no nucleus, typically there is a nucleus, surrounded by a viscous colloidal cell sap, bounded by an outer plasma membrane, and containing highly organized organelles such as mitochondria and systems of vacuoles. The cell sap is the only material substance to which the word "protoplasm" can properly be applied, though in this context as in every other it must be freed of its nature-philosophical connotation.

Organisms consisting of many cells are referred to as *metazoa* or *metaphytes*. The metazoa are sometimes thought to arise as colonies of individual cells—as clones of which the progeny, instead of separating, stick together. But a metazoan can equally well be thought of as an organism that has arisen by internal subdivision of a single preexisting cell, much as the fertilized egg of a frog or a sea urchin, say, turns into a many-celled organism. Protozoa are whole organisms whose structure corresponds to that of a single cell in a metazoan tissue. Whether they are described as single-celled or as "noncellular" is a matter of indifference to all except those nature-philosophers who have ensured their livelihood by inventing pseudoproblems and discussing them earnestly and at length, sometimes in specialist publications exclusively devoted to exercises of this kind.

The word "cell" has sometimes been used in the style of Robert Hooke (1625–1703), physicist, microscopist, and astronomer, who first discerned the cellular structure of cork. This did not mean, of course, that he believed that cork was composed of little holes; the existence of a structural boundary wall was implicit. We use the word in the same way when we refer to a cellular mattress, which is not thought to be composed solely of holes either.

The cell theory is the most important tectonic generalization in the whole of biology. Its most surprising and spectacular triumph is that embodied in the concept of the neurone, which is discussed in detail in the entry on the SYMPATHETIC NERVOUS SYSTEM. Propounded by Theodor Schwann (1810–1882), the theory declares that the tissues of animals are composed of cells. The other name we must mention here is that of Matthias Schleiden (1804–1881), the botanical microscopist who did for the cellular structure of plants what Schwann had done for animals.

CENTRIFUGE

When a system is composed of ingredients of different specific gravities, the elements of that system may be separated by means of an instrument known as a centrifuge. By increasing gravitational force, the centrifuge can quickly separate fat globules from milk, for example, or blood corpuscles from plasma—separations that can occur very slowly of their own accord by allowing the milk or blood to stand under ordinary conditions of gravity ($g = 1$).

A centrifuge for analytical or preparative purposes works essentially by spinning a bucket held in a suitable rotor and generating a centrifugal force (measured in gravitational units) akin to that which is produced when a stone attached to a piece of string is whirled around the head.

The rapid progress of technology is nowhere more evident than in the area of laboratory instrumentation—centrifuges with quite massive rotors may spin at speeds greater than one thousand times per second and are now quite commonplace laboratory equipment. For special research purposes airblown turbine centrifuges have been constructed to generate gravitational forces as high as 100,000 g.

CHANCE AND RANDOMNESS

Chance enters the fabric of natural processes at two levels relevant to the theory of evolution. First is the randomness in the occurrence and character of mutation and second is the built-in— one might almost say designed—randomness of the Mendelian process, the chance allocation of the members of pairs of chromosomes during the formation of gametes and their random conjunction at fertilization. Indeed, the randomness gives rise to an incredible variety of combinations—the trillions upon trillions of possible organisms that constitute the candidature for evolutionary change.

From the standpoint of the design argument and the presumed benevolence of a Designer, it will not do to count the blessings of randomness and turn a blind eye to its malefactions. Chance and randomness have built into the very fabric of nature the properties that make evolutionary change possible according to neo-Darwinian theory—and this must be judged a good mark. But there is a darker side, which is displayed in a most conspicuous form by reckoning the cost of fitness in a situation such as that which confers some degree of immunity to malaria upon the heterozygotes for hemoglobin S—in those that manifest sickle-cell trait. This relative immunity is fine for the heterozygotes, but rather hard on their progeny—a quarter of whom are victims of the virtually lethal *sickle-cell anemia.*

A theologian once put it to us that the wisdom and benevolence of the deity could be seen in the way in which the Designer provided for and made good use of the element of randomness as we see it in the Mendelian process, so that its outcome can be turned to profit. We cannot help feeling uneasy, however, at this use of an argument that would apply almost without qualification to the proprietor of a casino.

CHICKEN AND EGG

In the United Kingdom it is a long-standing tradition that even in the big cities milk is delivered daily to each household by a milkman, and often this is all that many townspeople know about its provenance. When London urchins were evacuated into the country at the beginning of the Second World War, many of them learned for the first time that milk was a product of the cow, but even those to whom this came as something of a revelation certainly understood that chickens hatched from eggs. This knowledge is so deeply rooted in the public consciousness that chickens and eggs have contributed over and over to the imagery and metaphors of science and of popular speech.

Consider, for example, the origin of one of the most famous epigrams featuring chickens and eggs. In his *Breeding and the Mendelian Discovery* (London, 1911) A. D. Darbyshire wrote as follows about the implications for biology of August Weismann's (1834–1914) doctrine of the complete separateness of soma and germ plasm:

Weismann's writings effected a swinging round of biological opinion through 180° which was the diametric opposite to that which had prevailed hitherto. The doctrine preached by Weismann was that to start with the body and inquire how its characters got into the germ was to view the sequence from the wrong end; the proper starting point was the germ and the real question was "How are the characters of an organism represented in the germ cell that produces it?" or, as Samuel Butler has it, the proper statement of relation between successive generations is . . . "to say that 'a hen is merely an egg's way of making another egg.' " (p. 187)

Darbyshire's attribution of the epigram to Butler is unequivocal, although Butler himself in a footnote to the second edition of *Life and Habit* (London, 1878) merely commented, "It has, I believe, often been remarked that 'a hen is only an egg's way of making another egg.' " There is not much pride of ownership here, but we believe that the reason for the form it takes is that Samuel Butler was too modest to say, as he might reasonably have done, "As I myself have so aptly and wittily observed, 'A hen is etc. . . .' " We must count it a mercy that the remark came too late to be attributed to Oscar Wilde. Butler's epigram, which embodies one of the most profound truths of biology—affirming as it does the primary importance of the replication of DNA—was unfamiliar to the late Jacques Monod, and we were fortunate enough to witness his delight upon first hearing it.

Hen and egg feature also in a famous riddle that is often thought to embody an unresolvable paradox: "Which came first, the chicken or the egg?" The two possible answers neatly define two entirely different biological and political philosophies. A person who believes that the egg came first—that a new kind of egg must have come before a new kind of chicken—is a Mendelist and a Morganist, and in the Western Hemisphere a trustworthy and regular guy (some compensation, perhaps, for the odium of being classified in the Soviet Union as a genetic elitist and a running dog of fascism intent upon still further subjugation of the working class). Conversely, someone who believes that the chicken came first—that a new kind of chicken, wiser or prettier, perhaps, than its parents, can produce a new kind of egg—would at one time in the Western Hemisphere have been denounced as a secret agent of the Comintern working to overthrow the constitution of the United States.

There are of course many other epigrams and tropes making use of the imagery of chicken and egg, though no other goes as deep as the two we have chosen: we stop short, for example, of contending that "Don't count your chickens before they're hatched" embodies some subtle demographic consciousness of the prevalence of infertility.

CHIMERA

Like much else in classical mythology, the chimera tries one's patience as sorely as it tries one's credulity. The classical chimera, the progeny of Typhon and Echidna, monsters both (evidently the condition ran in the family), was composed of three parts— a she-goat in the middle, with a lion before and a serpent behind.

In biological science, a chimera is an animal compounded of cells of two or more different genetic provenances. It is different from a *mosaic,* although a mosaic is also an animal containing cells of different genetic makeups; in a mosaic the genetic differences between the somatic cells have arisen by abnormalities of cell division affecting the chromosomes, especially in the early stages of development (thus some insects, gynandromorphs, are male on one side of the body and female on the other); or, alternatively, mutations may have appeared in one or another of the cell-division lineages that give rise to the adult body. Natural chimerism occurs regularly in fraternal twins in cattle, rarely in human beings. Twins of this sort often share a placenta and can thus commingle their bloods and blood-forming cells during fetal life. Such chimeras accept grafts from each other without rejection and thereby appear to flout the rules of transplantation—but this is true only because they become mutually tolerant of each other's tissues.

Twin chickens (two chicks hatching from one egg) may be either identical or fraternal; the latter are always chimeras.

Artificial chimeras can be manufactured by a laboratory reproduction of the phenomenon that occurs naturally in synchorial twins and can also be raised by fusing two very young embryos to form a single organism that has come to be described as "tetraparental." Another circumstance in which chimeras arise is when for any reason (for example, because of x-irradiation or gamma-irradiation) blood-forming cells have been destroyed and are replaced by an inoculum of blood-forming cells to make a genetically different animal. These are radiation chimeras. Chimeras comprised of elements as disparate as the classical chimera remain purely fictional beings: experimentalists do not try to produce them because such an enterprise would be uninformative and pointless.

CHORDATA

Chordata is the taxon that includes all vertebrate animals and many lesser groups related to vertebrates by evolutionary ancestry. Chordates are animals which at some stage of their life possess a *notochord,* a tough but flexible skeletal rod running along the dorsal length of the body. Until the evolution of paired limbs (something of an innovation), chordates typically moved by means of sigmoid snake-like contractions passing down the trunk, which straightens out by the elastic recovery of the notochord. In chordates the central nervous system is characteristically a hollow tube formed by the rising up of two as it were "shoulders" of the cells that constitute the outermost layer of the embryo, forming first a longitudinal dorsal gutter or deep groove and then, by a closure across the midline, a tube opening posteriorly. The heart is a ventral structure, and the principal body cavity, the coelom, is formed by the enlargement of a cleft that forms in the "mesoderm" between the outer ectoderm and the embryonic gut. It is characteristic of vertebrates that the anterior stretch of the gut (the pharynx) is perforated by clefts (gill slits) which serve a respiratory function, and that the somatic musculature originates in the form of segmental blocks lying on either side of the notochord and running from end to end of the body.

The closest living representatives of the primitive chordates that were probably ancestral to vertebrates are the thirty-odd species of lancelet that make up the genus *Branchiostoma* (= amphioxus), and this form in turn is very closely related by development to the animals known as *Tunicates* (sea squirts), the larvae of which have a notochord and a median dorsal nerve tube. Other animals too satisfy some of the criteria of the phylum Chordata, for example acorn worms and arrowworms, the word "worm" being improperly used in both contexts.

CHROMOSOMES

Chromosomes are the vectors of genetic information in all except the very simplest of the organisms, which are known accordingly as prokaryotes. Self-reproducing bodies of the nanometer order of size (one nanometer equals one billionth of a meter), chromosomes are housed within the nuclei of cells. They vary greatly in appearance, and their conventional characterization as threadlike does not describe them adequately; some are rather stubby and prickly, like caterpillars, and others more ribbonlike, as noodles are when compared with spaghetti. The complete confidence with which one can declare chromosomes the vectors of genetic information is built upon a variety of sources of evidence, of which the first and most important is the extremely exact correlation between the observed behavior of chromosomes and the postulated behavior of Mendel's "hereditary factors"; thus chromosomes are present in pairs, one member of each pair having been derived from one parent. In gamete formation chromosomes segregate, only one chromosome of each pair entering each gamete. Gametes accordingly end up with only half the parental number of chromosomes—with the haploid, as opposed to the normal diploid, number.

The reproduction of chromosomes is the primordial act of replication in biology: a chromosome is not synthesized de novo but is assembled as a copy of a preexisting chromosome; in other words, a chromosome forms only when there was one before. The same is true of many other cellular constituents such as mitochondria and chloroplasts, all described as self-reproducing. See our entry on NUCLEIC ACIDS.

In the union of gametes at fertilization the chromosomes pair off again and the diploid number is restored. When linkage (that is, inheritance of Mendelian factors not singly but en bloc) came to be known as a genetic phenomenon, it was satisfying to learn that the number of linkage groups was equal to the number of pairs of chromosomes, so that the chromosome could be regarded as a physical embodiment of the linkage group. Thomas Hunt Morgan (1866–1945), renowned American embryologist and geneticist, was the first to demonstrate that chromosomes are differentiated along their length. High-resolution microscopy eventually made it possible to identify various structural singularities along the length of the chromosome and often to correlate these with specific genetic effects; genes as such were not visible, however, being merely local combinational singularities of the nucleotides of nucleic acid.

In spite of the extremely powerful circumstantial evidence of the complicity of chromosomes in the hereditary process, some geneticists were unconvinced, or professed to be so. One of these was the English geneticist William Bateson, who in his great text *Mendel's Principles of Heredity* (Cambridge, 1909) did so much to make Mendelism known to the English-speaking world. It may be that Bateson was anxious for it to be more widely realized that the existence of chromosomes and of what later came to be called genes could be inferred from genetic evidence without recourse to the microscope. Bateson and others of like mind found many reasons for questioning the complacent assumption that chromosomes were the vehicles of Mendelian factors. Chromosomes are visible in the dividing cell, but as a rule not in the resting cell between divisions. It was at one time thought formally proper to express some misgivings about this apparent interruption, but the concern was wholly simulated; for in the light of genetic evidence positively requiring chromosomal continuity, no geneticist had any real misgivings. It is now known that some chromosomes *can* be seen in the resting nucleus (notably and most usefully the extra X chromosome of females—named the Barr body after its discoverer, Murray Barr).

Chromosomes are composed of deoxyribonucleoprotein—a salt-like compound between nucleic acid and a basic protein such as histone, the genetic function of which is not known, although the teleology of chromosomal organization is obvious enough. The chromosomes form a system of packaging nucleic acid in such a way as to prevent a variety of accidents which if they occurred would make nonsense of the essentially conservative process of Mendelian heredity and the orderly process of development (accidents such as the casual leakage of DNA into or out of nuclei, or the cross-infection of the nucleic acids of neighboring chromosomes). The chromosomes themselves do undergo accidents that have recognizable and sometimes very deleterious genetic effects; among them are multiplication of chromosome numbers above the normal or euploid condition. This is known as polyploidy, which sometimes takes the form of the presence in triplicate of a chromosome that ought to be present only in duplicate, as with the human chromosome 21 in Down's syndrome. Chromosomes also have a role in SEX DETERMINATION, which is elaborated in the genetic definitions.

Because physicists have played such an important part in the study of nucleic acids, it is worth remembering that it was a physicist, Erwin Schrödinger (1887–1961), who in *What Is Life?* (Cambridge, 1944) first described the chromosome as the physical embodiment of a code-script for heredity. The image became very familiar when the special relevance of information theory to biology came to be appreciated.

CLONES

A clone of cells comprises the progeny of a single parental cell, as a tumor is now thought to be, by repeated and successive mitotic (genetically symmetrical) cell divisions. Thus in a sense an adult organism is a clone descended from the zygote. The word "clone" may also refer to progeny of whole organisms that reproduce asexually. The connotation is that of identical replication. Thus identical twins that have been formed by the division into two equal parts of a fertilized egg constitute a first-order clone; indeed, the eggs of mammals are of the kind that lend themselves in principle to cloning, that is, to the production of a large if not indefinite number of replicates of a prototype. The practical realization of such a foolhardy ambition would call for the proliferation of a fertilized human egg under tissue culture conditions and the provision of an adequate supply of young foster mothers hormonally prepared for implantation of the egg. None of this is technically very difficult to accomplish: what is required all in one place is the ambition to accomplish it, a sufficient concentration of money, a modest technical proficiency, and folly.

In science fiction, however, and to some extent in popular speech, cloning has acquired an extra connotation: the replication of a fertilized egg which has been caused to acquire a genotype of choice by replacement of the zygote's own nucleus with the nucleus of a body cell of the person—a mad or vain millionaire, say—of whom replicates are to be made. An unscrupulous science journalist wrote a professedly documentary book on just such a theme, pretending it to be a true life story but at the same time declaring himself so bound by oaths of secrecy that there was no possibility of verifying the tale (or the probity of its author). The book was accompanied by a quasi-scientific appendix, and that was also a skillful exercise in the concealment of tracks. The authorities quoted were said to have been consulted pseudonymously by the author. Many people were taken in by this yarn, which we consider highly discreditable to author and publisher alike.

In our carefully considered opinion, cloning with this extra connotation of choice of genotype is not feasible in human beings. Cloning of plants is the phenomenon to which the word itself was first applied, with reference to the well-known propagation of plants by cuttings, runners, budding, or grafting; the term has now been extended to the raising of whole plants (for instance, whole carrot plants) from single cells. This usage resembles the others to which we have referred, in maintaining the essential connotation of asexual production of genetically identical replicates.

COMPARATIVE ANATOMY

In the great post-Darwinian era, when the principal business of biology was a demonstration of evolutionary descent ever more cogent and ever more specific, comparative anatomy was biology's central discipline. If carried out convincingly, comparative anatomy is an exact and an exacting pursuit that calls for the exercise of nice judgment and a certain amount of flair. Considered pedagogically, comparative anatomy has many of the virtues traditionally associated with the classical education called for by the humanities; it is understandable that in many eminent schools of biology it has remained the core of the teaching syllabus, one of the most famous of such schools being the department of zoology and comparative anatomy at Oxford University. Among its graduates have been one of the world's leading geneticists, a director of the London School of Economics, and a former director-general of UNESCO—so evidently the study of comparative anatomy does not unduly narrow the mind. The authors of this book write with a degree of pride excusable in alumni.

Among the triumphs of comparative anatomy is a striking example of the concept of homology. Comparative anatomists were the first to demonstrate that the miscellaneous animals formerly lumped together as "ungulates" walk on the tips of their toes—hence the name. The knee of the horse corresponds anatomically to the wrist or ankle; the horse walks upon the single digit corresponding to our middle finger or middle toe, digits 2 and 4 being reduced to "splint bones." The so-called even-toed ungulates such as sheep and cows walk on digits 3 and 4, with 1 and 5 being vestigial and sometimes disappearing.

Another interesting and unexpected finding in comparative anatomy was that the tiny bones of the ear that transmit vibrations from the eardrum to the organ of hearing were formed from the bones of the jaw liberated when the lower jaw of mammals was reduced to a single bone, the dentary. Again, the muscles that move the eyeball derive in evolution and development from the three most anterior muscle blocks of the series that at one time occupied the vertebrate body from end to end. In this sort of way, comparative anatomy has made evolutionary sense of the varieties of structure found among vertebrate animals.

CONTACT INHIBITION

The discovery by Michael Abercrombie (1912–1979) of a simple regulatory phenomenon, the mutual deterrence of amoeboid movements when cultivated cells come in contact with one another, explained much that had until then been mysterious (for example, the fact that the repair of artificial wounds inflicted on tissue culture proceeded only just far enough to make good the loss). Special interest was naturally aroused by Abercrombie's demonstration that contact inhibition is defective in malignant cells, a property that goes some way to account for the invasiveness of tumors. Future investigation of contact inhibition is likely to turn on defining, probably by immunologic methods, the structural properties of cell surfaces that conduce to contact inhibition—or under pathological conditions to its failure.

CORONARY ARTERY

The coronary is the great artery wrapped around the heart that provides the continuous and abundant supply of arterial blood demanded by the musculature of the heart (the myocardium). It is in a sense the most important artery in the body, because upon its correct functioning depends the functioning of all the others. Perhaps because of the continuous pounding and jolting it is exposed to, the coronary artery is rather particularly susceptible to the pathological changes to which arteries in general are vulnerable. Coronary arterial disease is frequently a precursor of coronary heart disease, which in extreme form may lead to complete occlusion, often followed by death of the heart muscle (myocardial infarction).

In the industrial Western world cardiovascular diseases have for some time been edging their way to the top of the list of causes of mortality.

CREATIONISM

Scientists other than biologists have noted with surprise, perhaps with annoyance, the bland self-assurance with which biologists have usurped the entire reference of the term "creationism." It is not surprising that they should have done so, because the biological is that aspect of creation upon which scientific evidence is most completely at variance with the biblical narrative—and the only aspect of it which had been the subject of a case at law. We refer to the notorious trial in Tennessee of a schoolteacher, John T. Scopes, for violating state law by teaching the theory of evolution. This trial is within the authors' memory, but is remembered by most others as the subject of a thrilling forensic confrontation between Spencer Tracy and Fredric March.

For biologists, then, creationism connotes the teaching that the Lord created every single species of animal and plant found in the world today, whereas biological orthodoxy has it that existing species arose over hundreds of millions of years as a result of the differentiation and divergence of their several biogenetic lineages—a hypothesis compatible with all the empirical evidence. The special creation of living organisms is part of the story, of course, but not the whole of it: what about those first verses of Genesis that answer the question "How did everything begin?" Surely creationism must comprehend the answer. The same goes for the origin of life, a subject upon which Darwin felt that speculation was useless.

We are surprised at the obstinacy with which creationists

cleave to literal creationism. So doing, they fail to realize that the evolutionary concept is a much grander and more awe-inspiring conception—in keeping with what C. S. Lewis referred to as rational piety, and for many people conducive to reverence. It is a pity that we still live under the dark influence of the single-handed inauguration by Bishop Samuel Wilberforce ("Soapy Sam") of an essential tension between biological science and religion.

Another consideration is that creationism in the biological sense—and especially such glosses upon it as Archbishop Ussher's (1581–1656) calculation that creation took place in 4004 B.C.— can be empirically faulted as myth. Norse mythology, for example, has it that the sun is pursued across the sky by two ravening wolves, Sköll and Hati, who cause partial or total eclipses of the sun by biting off part or swallowing it whole. The first few verses of Genesis cannot be empirically falsified and should perhaps be classified as imaginative literature rather than as myth. Purely scientific evidence has not much bearing on its acceptance: we cannot conceive of a frontier without conceiving that which lies on either side of it; how then can we conceive in scientific terms the frontier between being and nothingness, having in mind that nothingness can have no sensory content and cannot therefore be the subject of empirically based judgments? These are matters of faith: they lie outside science, and the popular idea that scientific evidence has refuted the biblical account of the primordial creation is not philosophically tenable.

CURARE

Curare is an arrow poison extracted by South American Indians from the root bark of certain poisonous plants, especially *Strychnos toxifera*. When introduced into the body parenterally (otherwise than by mouth), it blocks the connection between motor nerve and muscle. First to be paralyzed are the muscles moving the eyeball, then other voluntary muscles, and finally the muscles of the diaphragm; death subsequently occurs through respiratory failure.

Curare has abundant uses in surgery as a muscle relaxant. Neostigmine (an anticholinesterase) is kept at hand for use in emergency. The symptoms of poisoning by curare are in some ways similar to those of myasthenia gravis.

CYBERNETICS

This especially apt neologism was coined by Norbert Wiener (*Cybernetics* [New York, 1948]) from the Greek word *gubernētēs* (meaning "steersman") and intended by Wiener to refer to "the entire field of control and communication theory, whether in the machine or in the animal." In introducing this notion, Wiener makes special reference to a paper by James Clerk Maxwell published in the *Proceedings of the Royal Society of London* in 1868. Writing of governors, Maxwell gave an elementary mathematical treatment of positive and negative feedback and called attention to the inherent instability of the former.

In the exercise of control, the stratagem of *feedback* plays a particularly important part, for feedback is the control of performance by the consequences of the act performed. Of universal occurrence in biology, feedback is described as negative when the retroaction of output upon performance is to diminish it. An example in point would be the lowering of body temperature by the sweating that follows an elevation of body temperature. On the other hand, positive feedback is illustrated by the state of affairs that arises in certain "autoimmune" or immunologically self-destructive diseases, in which one of the first effects of the disease is to create anew and therefore reinforce the very conditions that led to the disease in the first place. An injury to the thyroid gland which gives rise to autoimmune thyroiditis (in effect, to antithyroid antibodies reacting upon the thyroid) leads to the formation of antibodies that generate still further damage to the thyroid and therefore acerbate the disease.

We have previously used a familiar everyday example to illustrate the principle of positive feedback: at a noisy cocktail party people have to speak louder and louder to make themselves heard. This adds to the prevailing din and so makes hearing even more difficult, causing everyone to speak louder still. Positive feedback creates an essentially unstable situation, leading in extreme cases to violent oscillations and even breakdown of the system.

It would not be feasible to enumerate all the situations in which the stratagem of feedback is used in the regulation of bodily performance, for feedback is ubiquitous. To give one example, in the endocrine system release of the hormones produced by the thyroid gland, the adrenal cortex, and the reproductive organs is stimulated by so-called trophic hormones produced by the pituitary gland. Production of these trophic hormones is switched off by the hormones whose liberation they stimulate, and thus the output of a hormone is maintained within physiological limits by the hormone itself.

As with information theory, the terminology and concepts of cybernetics were quickly taken up by biologists and have, so to speak, passed into the language—a sure sign that they were needed.

DEFINITION OF LIFE AND OTHER TERMS

In certain formal contexts—mathematical logic, for example, in which a definition is a rule for substituting one symbol for one or more others—definitions are crucially important, but in everyday life and in sciences such as biology their importance is highly exaggerated. It is simply not true that no discourse is possible unless all technical terms are precisely defined; if that were so, there would be no biology.

A principal purpose of definition is to bring peace of mind. Sometimes, though, it is too dearly bought: a "definition," as the word itself connotes, has a quality of finality that is often unjustified and misleading and may have the effect of confining the mind instead of liberating it. A great many nonbiologists believe that animated and contentious discussions of the definition of "life" are a principal preoccupation of institutes and university departments of biology. In reality, the subject is not mentioned at all, except perhaps to disparage the rather simple-minded people who believe that an agreed-upon definition of life will lead to a better comprehension of biology. Biologists already have a working understanding of "life" that is good enough for present purposes; we do not believe that any current research enterprise is at all impeded by the lack of a more formal definition. The trouble is that "life," like many other technical terms in science, has been pirated from the vernacular and is used in scientific contexts far removed from those that might arise in common speech.

Situations do certainly arise in real life in which a definition of "life" (or anyhow an exact description of all that "living" connotes) is genuinely important. Consider, for instance, the decision whether or not to use for grafting the kidney or other organ of a potential donor whose heart may still be beating. Such a decision turns upon a number of technical evaluations that belong to a world far removed from the entries in a sematological dictionary:

the assessment of brain function especially, and the question of whether the condition of the possible donor is reversible or not. These are factual, empirical questions that reference to a dictionary will not help to answer.

A hunger for definitions is very often a manifestation of a deep-seated belief—one of the many philosophical fancies made fun of in Lewis Carroll's enchanting satires *Alice in Wonderland* and *Through the Looking Glass*—that all words have an inner meaning that patient reflection and research will make clear and distinguish from false or counterfeit meanings, which may have usurped the true meanings; indeed, amateurs will sometimes put a question about definition in a form which reveals their enslavement to this illusion: "What is the *true* meaning of the word 'life'?" they ask. There is no true meaning. There is a *usage* that serves the purposes of working biologists well enough, and it is not the subject of altercation or dispute.

DEMOGRAPHY

The branch of learning that has to do with the growth, reproduction, and vital statistics of populations was first thrust into the forefront of public attention during the 1930s. Widespread fears that the people of the Western world were dying out through infertility before long were supplanted by the view that advances in medicine combined with the high fertility of the developing nations threatened an explosive increase in population, as disastrous in its way as decline through infertility.

In demography, as in developmental biology, a certain fastidiousness is called for in the use of technical terms. Every teacher who has heard students define "death rate" as "the number of people who die per thousand" would agree. We therefore include below a few definitions of frequently misused words and phrases.

DEATH RATE The notion of a death rate in demography sounds straightforward enough, but in using the notion one must have in the forefront of one's mind the nature of the independent variable in relation to which the rate is expressed. According to the uses to which it is to be put or the inferences that are to be drawn from it, a death rate may be expressed as the number of persons per thousand who die in a calendar year (for example, 1984) or in a year of age (in the seventieth year, for instance). Each has its special uses: in some ways the most informative measure is the *age-specific death rate*—the number who die in any year of age (between the years x and $x + 1$) expressed as a fraction of those still alive at age x. The age-specific death rate is often referred to as the force of mortality.

LIFE EXPECTANCY The registration of births and deaths that is commonplace in all civilized countries puts it within the power of a government's statistical office to estimate a person's average expectation of further life at birth or at any other chosen age. In principle, the calculation is quite straightforward: it involves estimating from the data the total number of person-years lived by people of any chosen age and dividing the product by the number of persons to live them.

The mean expectation of life at birth, often used as a measure of the well-being and general medical prowess of a population, has increased dramatically over the past few hundred years. This is true mainly because of the diminution of mortality in infancy and childhood (itself the consequence of the improvements in medicine and in sanitary engineering that have reduced deaths from infection to a very low level). The mean expectation of life at later ages has not increased nearly so much. Older people nevertheless have a lot going for them, not the least being the heightened resistance conferred by previous sublethal exposures to infectious agents—a sort of immunologic wisdom—and in addition by wisdom of a more conventional kind (embodied in the phrase "once bitten, twice shy").

Period	Sex	Mean expectation of life (in years) in Sweden at —		
		Birth	Age 60	Age 80
1755–1776	Male	33.20	12.24	4.27
	Female	35.70	13.08	4.47
1856–1860	Male	40.48	13.12	3.12
	Female	44.15	14.04	4.91
1936–1940	Male	64.30	16.35	5.25
	Female	66.90	17.19	5.49
1971–1975	Male	72.07	17.65	6.08
	Female	77.65	21.29	7.28

The mean expectation of life of females is at all ages—including perhaps from conception until birth—greater than that of males.

The most complete records we have are from Sweden. The accompanying table, which applies without difference of principle to North America and Great Britain, is seen to illustrate all the points that have been made above.

MALTHUSIAN PARAMETER   Ronald A. Fisher's term for Alfred J. Lotka's "true rate of natural increase" is an attempt to represent by a single figure the reproductive vitality of a population. Where b_x is the birthrate and l_x the likelihood of living to age x, the Malthusian parameter m is given by solution of the equation

$$\int_0^\infty e^{-mx} l_x b_x \, dx = 1.$$

The Malthusian parameter was thought by Lotka to be an improvement on the earlier net reproduction ratio (see next definition), but its use is open to the same objections.

NET REPRODUCTION RATIO   Robert Kuczynski made the first and most famous attempt to solve an intrinsically insoluble demographic problem: to express the reproductive well-being and the future growth prospects of a population by a single arithmetic figure—the net reproduction ratio, or NRR. (This is somewhat analogous to reading a patient's temperature on a clinical thermometer, as if it were possible to take the nation's temperature in the same sort of way.) The NRR may be viewed as an answer to the question, What is the likelihood that a specific individual today will be represented by a person of the same age in the next generation? If we confine our attention to female births only, the NRR may be expressed as the ratio of live female births in successive generations: if the value is unity, the population may be thought to be just holding its own; if less than unity, the population is reproductively losing ground; conversely, if the value is more than unity, it is increasing in number. Modern demographers do not accept the principle of measuring a nation's reproductive health in this way. The single-number valuation of complex quantities (see P. B. Medawar, *Pluto's Republic* [Oxford, 1982], pp. 167–193) is doomed to failure in this, as in many other contexts.

DESIGN, ARGUMENT FROM

The argument takes two forms, of which the more familiar might be described as the case from design for the *existence* of God. The argument is that Nature everywhere manifests a functional fitness and working capability that is evidence of design as clear as can be seen in a watch or engine or other such artifact. Design implies a designer, William Paley (1743–1805) reasoned; was this then not the supreme being, God? ("Why only one?" Hume had asked in his own skeptical reflections on the argument from design.)

The other form of the argument from design professes to infer not only the existence but also the *benevolence* of God. One does not have to be a card-carrying monadologist such as Leibniz (1646–1716) to conceive that the world might have taken a form other than the one in which we find it, but it would be downright impious to suppose that God, if He created the world, would create anything other than the best possible world. It is well known that Voltaire's *Candide* vigorously dissented from this view. Although in many ways a dull and tiresome book, it is, like so many satires, better known than the idea it satirizes—that this is the best of all possible worlds.

This second form of the argument is more ambitious: it purports to infer from the evidence of design not merely the existence of God, but some of His character traits also, especially His goodness. To demonstrate this was the ambition of the Right Honourable and Reverend Francis Henry Egerton, Earl of Bridgewater, who died in 1829 and directed in his will that the (at that time) munificent sum of £8,000 should be invested in gilt-edged securities so that the dividends therefrom might fund the publication of one thousand copies each of eight treatises "on the Power, Wisdom, and Goodness of God, as manifested in the Creation; illustrating such work by all reasonable arguments, as for instance the variety and formation of God's creatures in the animal, vegetable, and mineral kingdoms; the effect of digestion, and thereby of conversion; the construction of the hand of man, and an infinite variety of other arguments; as also by discoveries ancient and modern, in arts, sciences, and the whole extent of literature."

With the piety expected of—indeed statutorily required in—

the head of an Oxbridge college, William Whewell in his own Bridgewater treatise wondered whether "the whole mass of the earth from pole to pole and from circumference to centre" might not be "employed in keeping a snowdrop in the position most suited to the promotion of its vegetable health." This quotation, which we learned from Jonathan Howard's admirable *Darwin* (Oxford, 1982), epitomizes the spirit of the Bridgewater treatises—documents a modern Hume would make short work of. Another gem like Whewell's was exhumed by Sir George Porter (see his "Sir Humphrey Davy—Poet," *Interdisciplinary Science Reviews* 3 [1978]:262) from Davy's notebooks at the Royal Institution:

> The eternal laws
> Preserve one glorious wise design;
> Order amidst confusion flows,
> And all the system is divine.

In reality one can learn no more about the benevolence of God from the evidence of Nature than one could learn of the solvency of a company that presented its shareholders with a balance sheet containing credit entries only. And this is the philosophical gimmick of the Bridgewater treatises: Nature's success stories are lingered over and recounted, but the errors of nature, its many maladaptations, and the cost of fitness in blood, pain, and mortality are left unpriced. It is no wonder that God cuts such a good figure in the Bridgewater treatises.

DOWN'S SYNDROME

This is the medically approved name for the congenital affliction vulgarly known as mongolism, originally so called because of John L. H. Down's belief that it represented an example of atavism, an interpretation abetted by the supposed resemblance to the Mongolian facies of the broad, rather flat face and slanting eyes of some afflicted children. Although the mental retardation and general incapability that accompany Down's syndrome are its most distressing symptoms, the disease is marked by many other abnormalities of the internal organs combined with a general vulnerability to infections and the shocks of life. Although to some degree educable, children with Down's syndrome do not often become independent of special care and their life expectancy is lower than that of normal children.

The frequency of Down's syndrome in the general population, more than 1 per 1,000 live births, rises very steeply with the age of the mother. Obstetricians, therefore, now recommend amniocentesis for pregnant women over thirty and, a fortiori, for expectant mothers in their early forties, in whom the frequency of Down's syndrome may be more than 1 percent of live births.

Down's syndrome is a consequence of a gross abnormality of the chromosomes. Two classes of abnormality are especially associated with the affliction. One is *trisomy-21* (the presence in triplicate of a chromosome, number 21, that should be present only in the usual double form). The other abnormality is that in which chromosome 21 is present in excess because of the translocation to it of material from another chromosome. It is the trisomic form of Down's syndrome that increases in frequency with maternal age; Down's syndrome resulting from translocation is not so strongly influenced by maternal age, but unlike the other form may run in families.

Chromosomal abnormalities cannot be corrected. The treatment of Down's syndrome is confined to general care and supportive treatment, which in a humane society will extend to the parents of the afflicted child as well as to the proximate victim.

ECOLOGY

The words "ecology," "economics," and "ecumenism" all have their root in the Greek word *oikos,* meaning house or home. Ecology, topmost in the hierarchy of the life sciences, has indeed to do with the economy of the great house of nature, of which it seeks to reveal the structure in space and time and especially the interactions of animals and plants with themselves and with each other. Its content is enormous, for ecology enjoys the entire empirical content of the sciences below it in the hierarchy as well as, of course, the concepts contextually peculiar to itself.

The hierarchical ordering of the biosphere is discussed in the entries HIERARCHY and REDUCTIONISM. Just as characteristic is the cyclic ordering of the biosphere: cycles pervade the whole of animate nature. First, of course, we think of the slow pulse of the seasons upon which are superimposed the lesser cycles of the months—the tides and the alternation of night and day. The whole pattern is of cycles within cycles within cycles, the least of all—that with the fastest pulse rate—being the cell-division cycle in growth, in the continuing process of cell replacement and tissue repair, and in some physiological processes such as antibody formation. The movements of the elementary constituents of the biosphere from organism to and from environment, and from organism to organism, are also cyclic in nature. Carbon, oxygen, and nitrogen are never stationary, but are at all times on the move in the biosphere.

Thus in the food chain compounds of nitrogen pass from organism to organism, and nitrogen is eventually returned to the soil—or to the atmosphere by the putrefaction of living organisms and their excrement brought about by the bacteria that denitrify nitrogenous compounds and return molecular nitrogen to the atmosphere. Nitrogen is recaptured by microorganisms in the roots of leguminous plants that can use atmospheric nitrogen, and also by industrial processes on an enormous scale that convert atmospheric nitrogen into ammonia and nitrates which reenter living organisms through their use as fertilizers. These various events constitute the "nitrogen cycle," of immense importance because the cyclic movement of nitrogen is the limiting factor in what may be thought of as the "metabolism" of the biosphere (if we use that word in a sense not remotely different from its use in reference to individual organisms). Oxygen and carbon cycles are closely related and intersect in the process of photosynthesis, in which the energy of sunlight transforms water plus atmospheric carbon dioxide into carbohydrates, accompanied by the liberation of molecular oxygen, which is used in the respiration of both animals and plants.

Analysis of the cycles of the elements and preparation of balance sheets relating to them is a principal task of ecology. It is here that we naturally turn for guidance on such world problems as the conservation of water and the dangers entailed by the increase of carbon dioxide in the atmosphere resulting from the huge scale of industrial processes involving combustion and therefore consumption of oxygen. More carbon has been locked up in the form of coal and other fossil fuels than in the living organisms of the biosphere. Plants, of course, make use of atmospheric carbon dioxide, and again it is to ecology that we look in attempting to assess and remedy the consequences of the destruction of forest lands. From the standpoint of the metabolism of the biosphere the participation of the sea, and especially of the minute marine plants of surface waters (the phytoplankton), are quantitatively of the highest importance.

Thus ecology, meteorology, and geophysics have many common interests. Modern ecology has moved a long way from the days when a biologist with a vasculum, lens, butterfly net, and supply of sandwiches rated himself an ecologist if he recorded the varieties and the degree of abundance of living things in a freshwater pond, a rock pool, a copse, or an area of arable land. Everything is now very sophisticated and depends upon the use of techniques no longer within reach of the amateur. Moving animals may be made to declare their presence by minute radio transmitters that betray their positions: worldwide schemes of marking and recapture uncover major migrations of animals and are invaluable aids in the extraordinarily difficult task of compiling life tables for living organisms in a natural environment. The concept "natural" is crucial, for ecology has in common with ethology a strong conviction that its principal business is with the natural world—with the world as it is, rather than with experimental simulations of it. Even though catastrophes and vast movements of land and water occur that may be thought of as "natural experiments," ecology is not an experimental subject in a Baconian sense: there is no need to contrive situations that do not occur naturally in the way a chemist is obliged to contrive conjunctions of materials or of events that would not come about under natural circumstances. It is a matter of particular satisfaction that ecology now enjoys enough political clout to insure it against languishing, as do so many enterprises which are thought not to assist the advancement of politicians or of the parties for which they stand.

ELECTROPHORESIS

Biological macromolecules as a rule carry an electric charge, positive or negative as the case may be, and by reason of this property will move in solution toward one or the other pole of an applied electric field. Electrophoresis thus does for colloidal particles what electrolysis does for inorganic solutes that undergo ionization in solution. The technique, introduced by Arno Tiselius, has been of immense help in characterizing and isolating biological macromolecules. An especially useful variant is to bring about electrophoresis in a gel rather than in a liquid: it is then possible to separate various classes of compounds by using a sharp blade to cut the gel to lengths, one of which may contain the compound sought.

EMBRYONIC AXIS

In the course of their development all vertebrate animals go through an embryonic period known as the neurula, a stage in which the larva has many of the characteristics of a prototypical chordate. The embryo is an elongated, bilaterally symmetrical organism, somewhat flattened from side to side, with a well-defined head bearing paired sense organs, and a spacious body cavity (the coelom) formed as a split within the cells that make up the intermediate (mesodermal) layer of the embryo. The kidney is shaped from the stretch of coelom connecting the cavities of the segmental muscle blocks with the main perivisceral cavity ventral to it. The formation of the nerve tube is described in the entry CHORDATA. The axial structures are of course the nerve tube and the notochord.

ENDOCRINE GLAND

The most familiar kind of gland empties its secretion into a duct that passes from the gland to wherever the secretory product works. The parotid, a salivary gland, is one of these "exocrine" glands, its secretion passing through a duct into the mouth.

The secretions of certain other glands (the adrenal, for example) act systemically and not just locally. Such glands—the "endocrines"—are ductless and empty their secretions directly into the blood or lymph. Several glands are both exocrine and endocrine: in addition to producing digestive juices, the so-called islet cells embedded in the pancreas manufacture the hormone insulin and secrete it into the bloodstream; likewise, the gonads produce germ cells and also internal secretions affecting sexual development and behavior.

The evolutionary developments of the several endocrine glands have a number of features in common, one of the most striking being that a number of endocrine glands are homologues, evolutionary descendants of organs that have become obsolete. The thyroid gland, for example, which is so vital to the regulation of metabolic rate, is the evolutionary descendant in modern land vertebrates of an organ that at one time was part of the food-gathering apparatus of the most primitive chordates—a mucus-secreting groove or gutter in the floor of the pharynx, which served to spread across the inside of the pharynx the sticky secretion that entraps minute food particles passing through the pharynx in the respiratory stream. Again, the all-important anterior element of the pituitary gland (important because other endocrine glands are among its target organs) was at one stage in evolution a functional median nose that passed from the roof of the mouth to the floor of the brain.

How did evolutionary relics come to acquire a significant endocrine function?

It is most unlikely that endocrine function was acquired de novo: it is much more probable that the organ simply retained an endocrine function it already possessed when its primary function became redundant. It is after all very likely that many organs have an endocrine function related to their principal physiological activity (as the ovaries do, for example, and the testes); thus the pharangeal feeding groove referred to above as the evolutionary progenitor of the thyroid gland concentrates iodine as the thyroid does, and almost certainly manufactures a hormone of the thyroidal type.

The classic way (by no means the only one) of demonstrating endocrine function is to extirpate the gland, observe the consequences, and then demonstrate that normality can be restored by replacement of the gland or by injection of an extract made from it. The procedure is of course not feasible when the gland cannot be extirpated, either for anatomic reasons or because in its everyday capacity it is essential for continued life. This applies with special force to the liver (very probably the source of a whole variety of hormones) and to the brain (part of which, the hypothalamus, is responsible for secreting the hormones formerly supposed to be synthesized in the posterior part of the pituitary gland). In cases such as these, circumstantial evidence provided by microanatomy, histochemistry, and the like may raise a strong presumption of endocrine function even if it cannot be proved with the elegant finality of the procedures that disclosed endocrine function in the adrenal gland, the thyroid, and the pancreas.

It is in accord with our theory of the retention of endocrine function by organs that become functionally obsolete in the course of evolution that the "modern" endocrine glands do very often exercise a function related to the original function of the organ. Thus the medulla of the adrenal gland exercises much the kind of function one would expect to find in an evolutionary descendant of a sympathetic ganglion, and the regulation of body salts is a function not unexpected of an organ—the adrenal cortex—that may represent the obsolete *pronephros,* the first-formed vertebrate kidney.

Another curious and unexplained characteristic of endocrine evolution is how frequently it has happened that two endocrine organs of completely different origins and functions have come together in the course of evolution to form a single composite organ in the manner of the cortex and medulla of the adrenal gland, the thyroids and parathyroids, and the anterior and posterior elements of the pituitary gland.

ENZYMES

Most of the molecular transformations that constitute metabolism are mediated through the action of proteins known as enzymes that expedite such transformations so greatly that for all practical purposes they could be judged not to occur at all in the absence of the enzyme. Many enzymes depend for their action on the cooperation of accessory factors, sometimes referred to as coenzymes, which often have enzymic properties themselves. Trypsinogen, for instance, precursor of the most important protein-splitting enzyme of the small intestine, trypsin, is activated by the enzyme enterokinase. Enzymes by and large are specific in their interaction with substrate and are very exacting with respect to the temperature and degree of acidity or alkalinity at which they work best. Thus the most important protein-splitting enzyme of the stomach, pepsin, works in a quite strongly acid medium and hardly at all in an alkaline medium.

The molecule upon which an enzyme acts is referred to as its substrate, and the accepted nomenclature for enzymes is formed by adding the suffix *-ase* to the substrate; thus proteases, lipases, and carbohydrases act respectively upon proteins, fats, and carbohydrates. The term "proteolytic enzyme" for a protease is obsolescent, and "proteoclastic enzyme" is already obsolete. As with all other proteins, the assembly of enzymes is specified by the nucleic acids that encode the information directing their synthesis. It can hardly be doubted that the specification of enzyme synthesis is a principal function of the nucleic acids transmitted in the germ cells. Mutations that affect the synthesis of a particular enzyme will give rise to the corresponding enzyme-deficiency disease. We have chosen phenylketonuria as an example: it is caused by a gene which when inherited from both parents leads to a deficiency of phenylalanine hydroxylase. In albinism the deficient enzyme is tyrosinase.

Enzymes also have a wide range of uses in biological manufactures, as discussed in our entry on BIOENGINEERING.

EPITHELIUM

The epithelium is a tissue composed of contiguous cells, all of the same kind, so arranged as to bound a surface. In its original usage an epithelium bounded a convex surface, whereas an *endothelium* bounded an inner or concave surface. Thus the cornea, the window of the eye, is lined on the outside by an epithelium and immediately behind it the anterior chamber of the eye is lined internally by an endothelium.

One of the very few respects in which we have ever differed with the *Penguin Dictionary of Biology* is that we regard the term "mesothelium" as unnecessary and etymologically dubious. (We presume it was so called to distinguish an epithelium of mesodermal origin—a quite unnecessary intrusion of the germ layer theory into anatomy.) The cells that form the internal lining of the heart and blood vessels throughout the body are always referred to without fear of confusion as the "vascular endothelium."

ERRORS OF NATURE

The following experiment, which we would prefer had not been done, was carried out on rats deprived of their adrenal glands.

The cortex of the adrenal gland is essential for life, and its removal or inactivation leads to a rapid loss of salt from the body, with fatal results unless a compensatory excess of salt is eaten. When rats deprived of their adrenal glands were offered a choice between plain and salty drinking water, they drank the latter and survived; but when the choice was between salty water and sugary water, they drank the sugar water and died. The moral is that Nature does not always know best and that natural appetites may mislead. The Arcadian notion that Nature does know best is widely prevalent; it underlies, for example, the herbalists' pious belief that Nature is designed to provide from its own bounteous pharmacopoeia of berries, worts, and roots a remedy for all natural human affections.

Nature must not, of course, be blamed for our misconceptions of it any more than it can be blamed for the miscarriages of such artificial procedures as blood transfusion and tissue transplantation between genetically unrelated individuals. Those who persist in their advocacy of the argument from design have a valid case to answer; they must meet an accusation just as well founded as that repeatedly leveled against doctors whose patients suffer from the side effects of drugs or from any other misadventure of supposedly therapeutic procedures.

Many natural adaptations are strongly contra-adaptive because of side effects that make people ill or make them more susceptible to illness than they otherwise would have been. Consider for example those prevalent sources of human distress, the allergies, which in their various manifestations—pollen or food allergy, anaphylaxis, the self-destructive autoimmune diseases—are all indirectly a consequence of our enjoying an immunologic response system so extremely sensitive and so very well able to discern the intrusion into the body of nonself substances: no immunologist would question for a moment the proposition that a rabbit not yet born will be able to manufacture an antibody against a chemical compound not yet synthesized. Of course, it may properly be said that the possession of an immunologic response system confers upon us and upon animals generally a *net* advantage—for such a system is essential to life. Although a man or woman prostrated by an asthmatic seizure might think it a life not always worth living.

A second, different example of a maladaptive side effect is that associated with human upright posture, which, it has been remarked, may be a constant source of moral satisfaction but has serious mechanical drawbacks. In most mammals the backbone is a simple uninflected arc running from head to tail—a cantilever having the four legs as piers—but the upright human backbone has several points of inflection: it bends somewhat forward in the thoracic cage and in the small of the back and somewhat backward in the sacrum. Upright posture exposes the backbone to new and evolutionarily unfamiliar physical shocks and stresses, so it is no wonder that the spinal column has the reputation of being the first structure in the body to show signs of "aging." The great apes, though facultatively two-legged, avoid many of the disadvantages of that condition by using their arms for walking and their knuckles as forefeet, walking characteristically not on the flats of the feet but on their outer margins. Human beings, walking upright and being squarely flat-footed, are extremely vulnerable to such ailments as the anterior protrusion of intervertebral discs and a variety of back pains of uncertain etiology.

A veterinarian once was deeply enraged by the description of our vulnerability to injuries of the spine as an "imperfection of man": dogs, he insisted, also suffered from disc lesions. It is unfortunate that he so badly missed the point of this description of a disc lesion (as the maleficent side effect of an adaptation) that he dismissed the whole argument as yet another conspiracy to diminish the standing of veterinarians and to deprive them of self-respect arising from the fact that their patients suffer ailments of the same kind and degree of gravity as those that afflict human beings. He was probably not appeased when, in correspondence, he was invited to draw a distinction between the maladaptations of otherwise normal human beings and those that may afflict the mutant horrors which form so large a part of small-animal veterinary practice.

We must not leave this subject without stressing that the net advantages of the upright posture strongly outweigh the disabilities of the kind we have mentioned: for the upright posture liberates the hands and so makes dexterity possible—a priceless possession.

A miscarriage of immunity underlies a third example of maladaptation: the susceptibility of human beings to hemolytic disease of the newborn. This affliction arises because the placenta is so attenuated as to be permeable by the antiembryo antibodies that are sometimes formed (see our entry on RHESUS FACTOR) when at parturition fetal red blood corpuscles escape into the maternal circulation. Rhesus polymorphism may of course confer some advantage that more than balances the wastage from hemolytic disease. It is hard to see what that advantage is, because it is not present in anthropoid apes nor in all human communities.

The fourth example is of a maladaptation of an altogether different kind: a state of affairs in which an adaptation that is clearly advantageous for a majority of people is fatally disadvantageous to a few. In the entry SICKLE-CELL ANEMIA we explain that sickle-cell trait is the heterozygous manifestation of a gene S, which transforms normal hemoglobin A into an abnormal form that causes the red blood corpuscles to become sickle shaped in deoxygenated blood. Sickle-cell trait is widely prevalent in areas of the world where malaria is or has been endemic, because its victims enjoy a relative immunity from malaria. This adaptive coup appears on the surface to be what a chess player would call a "brilliancy"—a perfect example of the wisdom of Nature and the beneficence of its designs. But alas, there is a debit entry too, of just the sort that the author of a Bridgewater treatise would be well advised to disregard: approximately one quarter of the offspring of parents showing the sickle-cell trait will die young, because in accordance with Mendelian rules this is the fraction that will be homozygous for gene S and will suffer from sickle-cell anemia, a killing disease.

These are all cautionary tales: they admonish against imputing to nature a design that would tempt one to infer the existence of an omniscient Designer who is able to foresee the injurious side effects of His arrangements.

ETHOLOGY

In the nineteen twenties and thirties the study of animal behavior—or of that part of it which its practitioners would have deemed to be "scientific"—was in a sorry state. Living organisms were relegated to the position of laboratory performers, and experimentation took the form of ascertaining and recording the way in which an episode of behavior was influenced by modifying the conditions under which it was performed. To observe an animal was not enough: the animal had to be pricked, poked, prodded, or have a bright light shone into its eyes, whereupon its behavior was modified in a way that was automatically thought to be informative (though in reality it seldom was). Merely to observe an animal, especially in its natural environment, was not rated a scientific activity but rather an agreeable outdoor pastime, suitable for clergymen on weekdays and for enthusiastic young mothers inculcating habits of intelligent awareness into their progeny.

The word "ethology" is not merely an alternative designation of the science of behavior: it is a term that stands for a genuine revolution in biological thought. Ethology is rooted in the *observation* of animal behavior, an activity that only simpletons think simple. Except in the special case where an observation merely confirms or rebuts a prior expectation, observation is a difficult and sophisticated process calling upon all the intellectual virtues: attention, patience, heightened awareness, caution in coming to conclusions, courage in framing expectations. Ethologists characteristically study natural as opposed to contrived behavior and try also to discern the *structure* of natural behavior, that is to say, to discern a functionally coherent, quasi-purposive performance in what to an inexperienced observer would present itself as a sequence of isolated and teleologically unconnected performances.

It is particularly satisfying to record that the first behavioral study that would be recognized by any modern ethologist as ethological in spirit and in execution was the work of an all-around experimental biologist who had made important contributions to genetics, embryology, and the study of growth. We refer to Julian S. Huxley's paper on "The Courtship Habits of the Great Crested Grebe," delivered to the Zoological Society of London on April 21, 1914, and published in volume 35 of the society's *Proceedings*. Study of animal behavior with the intentness that will discern a behavior structure is by no means a matter of passive observation: it is an exploratory process in which the observer is constantly framing hypotheses, forming expectations about what is going on—expectations which repeated observations will either uphold or rebut. By such means ethologists have built up an impressive anatomy of normal behavior. Behavior is studied under natural conditions because the interpretation of natural behavior is the goal of ethology. Experimental interference—such as the addition or removal of one egg from a nest—is undertaken only insofar as it will throw light on natural behavior. There is no experimentation for its own sake to create the appearance of being laudably scientific.

Ethology has been more successful on the descriptive than on the causal/explanatory side—that is to say, in the interpretation of behavior in terms of the electric traffic of brain and nervous system, which is not the interpretive level at which ethologists choose to work. Learning and memory do not therefore rank as high on the agenda of ethology as they did on that of the older style of investigation which ethology has largely supplanted. What is happening is that ethology is building up a library of contextually distinctive concepts such as drive, appetitive behavior, displacement activity, and the like, in terms of which behavior can be explained at an ethological level much as political behavior can be explained at a political level without recourse to psychological interpretation.

If there is human physiology, human anatomy, and human genetics, why should there not also be human ethology? No reason at all. It simply has not yet been attempted in any systematic way. When the connection between smoking and lung cancer came to be known, it would have been helpful indeed if ethologists had already studied the ethology of smoking and categorized its various forms more rigorously than the familiar but inadequate distinction between inhalers and noninhalers. It is a job that should be professionally done, though of course we are all familiar from our own experience with the medically important distinction between, on the one hand, those who take occasional drags at a cigarette (leaving it between puffs in an ashtray if they have been well brought up or, if not, over the edge of a priceless marquetry table) and, on the other hand, those whose cigarettes bob up and down in the mouth as they talk (while their eyes smart and blink at the unfiltered smoke that coils upward from the lighted end). Those who have given up smoking feel like people freed from a tyranny, especially from such appetitive activities as the patting of pockets, the groping in handbags, and the opening and shutting of drawers that may eventually confirm the dreadful suspicion that they have run out of cigarettes.

The "Olympian glibness" of psychoanalytic interpretations of human behavior have often been held up to public obloquy (see "Further Comments on Psychoanalysis" in P. B. Medawar, *Pluto's Republic* [Oxford, 1982], pp. 62–72). But ethology's excursions into the interpretation of human behavior have been just as glib, and in some ways even more culpable because there is a kind of plausibility about them which has often taken in the inexperienced or the unwary. We have in mind quasi-ethological interpretations of the origin of human aggression, tribalism, and territorial possessiveness. The most confidently self-assured of these are something of an embarrassment to old pros among ethologists, who often firmly dissociate themselves from such theories. This is not to say, of course, that the study of animal behavior throws no light upon human behavior: it would be an inexplicable anomaly if it did not, for human beings evolved from lower animals and much of their behavioral repertoire is homologous with that of animals. Maternal care is a case in point, for no one in his right mind could suppose that the entire complex repertoire of maternal behavior in human beings arose de novo with the evolution of man. This is a simple example, of course, because there is so much independent and parallel evidence—from embryology, endocrinology, and elsewhere—that upholds the homology. From the standpoint that would have been taken by a modern Dr. Johnson or Voltaire, what is most interesting about the behavior of man is not the part that can be construed as an inheritance from lower animals but, on the contrary, all the greater and more important part that comprehends distinctively human practices and propensities such as moral judgment and ratiocination.

It is in any event reassuring to know that, far from being distinctively human, play, aggressiveness, bluff, and showing off are part of a very ancient behavioral heritage. Nor is there any human diminishment in recognizing, as did Darwin in his *Descent of Man* (London, 1890), that man still bears in his behavioral repertoire "the indelible stamp of his lowly origin."

The descriptive ethological concepts now to be outlined arose in response to a need and have persisted because they satisfied it. Though many are figurative in nature (a characteristic not to be avoided in the English language) the terminology is natural and unpretentious and can be understood easily by those who wish to do so. *Drive,* for example, is a figure of speech for the *motivation* (another figure of speech) which underlies the behavior that satisfies certain basic biological needs; expressions such as "sex drive" and "hunger drive" have an immediate intuitive intelligibility. A drive tends to persist until its goal, a *consummatory act,* is achieved. Thus the hunger drive is, as it were, switched off by eating and the sexual drive by mating. The kind of behavior that increases the likelihood of achieving such a goal is described as *appetitive behavior,* something that applies literally in the food drive and figuratively in the sexual drive. It has been a commonplace observation of ethologists that if a drive is frustrated or if an animal is confronted with incompatible alternatives—as happened with the proverbial ass that died of hunger because of its indecision in choosing between two equidistant and equally large bundles of hay—then an animal may distract itself by a flurry of activity of no obvious functional significance. This is a *displacement activity,* such as a scientist might be thought to indulge in when, stuck in the composition of a paper intended for a learned journal, he builds a useless piece of apparatus or carries out a deeply uninformative experiment.

Doctrinaire behaviorists are known to regard this entire special terminology as an affront. Persons of more temperate judgment will instead rejoice that the richest and most versatile of all languages has the resources for building an apt and unpretentious terminology so well suited to its purpose.

EUGENICS

Most truthfully described as a body of aspirations having ostensibly to do with the genetic welfare of mankind and resting upon the authority of a science (genetics) imperfectly understood, eugenics is now so mixed up with racism that it is difficult to appraise it coolly. Yet an attempt to do so is imperative if we are not to continue at the mercy of ambitious politicians who might once again abuse the authority of genetics to promote their mischievous and illiberal ambitions.

The word (for which he at one time used "viriculture") was coined by Francis Galton (1822–1911), who in 1908 introduced the notion in the following terms: "Man is gifted with pity and other kindly feelings; he has also the power of preventing many kinds of suffering. I conceive it to fall well within his province to replace Natural Selection by other processes that are more merciful and not less effective. This is precisely the aim of eugenics."

These words, if murmured while gazing at a photograph of Galton in his prime—a benevolent-looking gentleman with the air of an elderly London clubman—might create the impression that Galton, deep at heart, was a reasonable and kindly scholar working to spare mankind the cutting edge of natural selection. Any such interpretation would be wide of the mark indeed: Galton envisaged the development of a strong "caste sense" among the naturally gifted members of each social class. These, the genetic elite, would have the biggest say in legislation and, where called for, the first claim upon charity. Galton wrote: "I do not see why any insolence of caste should prevent the gifted class, when they had the power, from treating their compatriots with all kindness, so long as they maintained celibacy. But if these continued to procreate children inferior in moral, intellectual and physical qualities, it is easy to believe the time may come when such persons would be considered as enemies to the State, and to have forfeited all claims to kindness" (*Fraser's Magazine* 7 [1873]).

So much for all that fine talk about man's being "gifted with pity and other kindly feelings." Galton's is the morality of the gas chamber. The worldly-wise will not think it at all out of character that Galton had an unhealthy interest in the subject of capital punishment, especially in death by hanging (*Nature* 57 [1897]:79).

The supposedly scientific justification of eugenics is a parallel or analogy which we shall show to be deeply unsound. It is the practice of the stockbreeder: if the stockbreeder can improve his stock by selective mating, why cannot viriculture do as much for man? The first riposte that comes to mind is that selective breeding is not the stockbreeder's only stratagem: he has recourse also to *culling,* and this is something that no eugenicist has had the temerity to propose for man—unless perhaps Galton believed that the same effect would be achieved by the withholding of charity, and perhaps of medical attention, and of anything else entailed by the forfeiture of all claims to kindness, to use Galton's own unhappy expression.

The technical objections to the parallel with stockbreeding, however, are more damning scientifically. The elite caste that Galton envisaged would not be of much use unless its progeny had the same superior characteristics—something which, it was confidently thought, could be achieved by bringing it about that members married only each other, because to do otherwise would be to undo the selective work of many generations of controlled matings.

The genetic implication here accords perfectly with elementary genetic notions that had prevailed for many years, namely that the well-adapted animal was homozygous with respect to the genetic factors being bred for, and that animals homozygous for certain genes mated with others homozygous for the same genes would breed true and produce progeny genetically similar to each other and to themselves with respect to those genes. The elite, then, would be homozygous with respect to the genetic factors that made them so. Unfortunately, this is far removed from genetic reality. In nature populations are not homozygous but quite extensively heterozygous, although the pattern of genetic inequality in the population may itself be fairly stable—lessons learned from the observations of Th. Dobzhansky, E. Mayr, and many other population geneticists of distinction. The elite thus could not be relied upon to breed true, an ambition that could be fulfilled only if the genetic factors that made for eliteness were so few that selective breeding could "fix" them in the breeding stock in the homozygous state.

So far is it from true that the end product of the stockbreeder's art is a homozygous population, that stockbreeding is coming increasingly to rely upon production and marketing of predominantly heterozygous end products, themselves the progeny of a cross between two parental stocks neither of which need be homozygous. Indeed, the idea that a superior caste of human beings can be raised up and will perpetuate itself by intermarriage has no solid genetic foundation: it could be done, of course, if we were content with the fixation of one or two genes thought to be specially desirable, but the ambition of eugenicists goes well beyond such modest procedures.

The notions thus far criticized do not relate to the program that is sometimes called negative eugenics—the putative improvement of the human genotype by piecemeal genetic intervention, especially that which makes use of marriage guidance and genetic counseling, without any element of coercion or threatened withdrawal of privilege. Such measures are preventive: their purpose is, insofar as possible, to prevent any foreseeable conjunction of genes that might cripple a child physically or biochemically. Although some abnormalities respond to this approach (among them phenylketonuria, hemophilia, and Huntington's chorea) congenital abnormalities that are due to abnormalities of the chromosomes (Down's syndrome is the most obtrusive example) form a category apart. There is no doubt that the frequency in the population of Down's syndrome has been dramatically reduced by a turn of opinion that decreased the mean age of motherhood by five years.

Because abnormalities of the chromosomes and the diseases they give rise to are virtually irremediable, afflictions of this kind raise all the problems and threats that pertain to feticide, but these are not the province of the geneticist. His role is, in collaboration with the physician, to draw attention to the facts of the matter and give grounds for forming an opinion on whether or not premeditated feticide is a desirable course of action. Such a judgment would doubtless be formed in the clear realization that no one has conferred upon—or indeed withheld from—human beings their supposed right to bring physically or mentally disadvantaged children into the world. There is, to be sure, sociological evidence which makes it seem likely that modern parents will not stand on any hypothetical right to bring defective children into the world. We can cite, for example, the popular reaction to the discovery that hemolytic disease of the newborn is most often due to incompatibility of rhesus factor; because of the prevalent misconception that *all* children born of rhesus-positive fathers and rhesus-negative mothers will be afflicted by hemolytic disease (which is by no means the case), young couples who had been blood-grouped sometimes broke off their engagements.

We shall now give a number of examples of the forms which genetic counseling might take in order to circumvent genetic disorders. Consider first the case of hemophilia, an exceedingly painful and disabling abnormality of blood clotting that makes its victim a "bleeder." In the most familiar, most thoroughly studied form, it is a sex-linked recessive condition in which the victims are mainly male, although the offending gene is carried by females: the working of the Mendelian rules is such that half the sons of a female carrier will be afflicted and half the daughters will be carriers. In such a case the only wise and charitable procedure is for the woman carrier of the gene to abstain from childbearing.

For somewhat different reasons, an analogous procedure should be adopted with the grave abnormalities that are genetically controlled by late-acting dominant genes. Huntington's chorea is one such disease, and because its effects become manifest somewhat late in life, a future victim may have children before the disease is evident. On average, half the children of a marriage between a normal person and a future choreic will be carriers of the dominant gene and will be afflicted; the other half will be ostensibly normal. A choreic who marries and has children thus will bring into the world offspring who will either contract a most terrible disease late in life or live in the shadow of a dreadful fear that they will eventually do so. Here too, the only humane and socially just procedure is to abstain from having children.

The so-called recessive diseases include phenylketonuria, a principal cause of low-grade mental deficiency. While it is most unlikely that all such diseases have yet been identified, the order of frequency of those that are known is 10^{-4} (one in tens of thousands).

These conditions are described as recessive because the gene responsible must be carried by *both* the gametes that unite to form the new individual. In view of the infrequency of mutation and the infertility of phenylketonurics, the great majority of the afflicted will be the offspring of marriages between two heterozygotes, that is, two carriers of the same offending gene. On the average, one-quarter of the offspring of such a union will be normal, one-quarter afflicted, and one-half carriers like their parents. In the table below we have taken two illustrative frequencies of afflicted persons: one in ten thousand and one in forty thousand. By application of the Hardy-Weinberg Theorem, the frequencies of the heterozygous carriers of the gene will be one in one hundred and one in two hundred, respectively.

Frequency of afflicted — p^2	1:10,000	1:40,000
Frequency of carriers (heterozygotes) — $2p$	1:50	1:100
Frequency of marriages between carriers — $(2p)^2$	1:2,500	1:10,000
Gene frequency — p	1:100	1:200

It is clear that if these carriers can be identified (something possible in an increasing number of recessive disorders, but not yet in all), then the appropriate preventive measure is to discourage marriage between carriers of the *same* harmful recessive gene. An eminent anthropologist, anxious to find fault, construed this recommendation as the truly ludicrous proposal that marriage should be discouraged between carriers of harmful recessive genes (a proposal which, if put into effect, would effectively abolish the institution of marriage altogether because *everyone* is a carrier of one or more harmful recessive genes). If the frequency of carriers of the two conditions we are considering is one in fifty and one in a hundred respectively, then, as the table shows, the proportion of marriages discouraged would be one in twenty-five hundred and one in ten thousand, figures that cannot reasonably be construed as a serious threat to the liberty of the subject or a wantonly cruel deprivation of a person's supposed birthright to bring physically or mentally disabled children into the world.

Nevertheless, the procedure is not as simple as it may appear at first sight. It would entail some screening of engaged couples, which could only be done on a voluntary basis and only when family histories raised the likelihood or even the remote possibility that one or the other parent might be a carrier. Genetically, moreover, the withholding of natural selection can only have the effect of compounding the problem; for although it may be masked in the heterozygous state, the offending gene will continue to increase in frequency through the ordinary pressure of mutation. Whatever problem the gene currently gives rise to will become progressively worse generation by generation. It is not much consolation that before the situation becomes unmanageable, a cure for phenylketonuria will be found; for the cure will not be a genetic cure, but only a symptomatic cure which, like the discouragement of intermarriage between carriers, could only be dysgenic in effect.

The problems of how to cope with genetic disorders are grave and exigent and will become more so as mortality from other causes diminishes. But solutions can be found if tact, humanity, and wisdom are applied. See *Francis Galton: The Life and Work of a Victorian Genius* by D. W. Forrest (London, 1974).

EUKARYOTES

This term refers to cells, or to organisms containing cells, in the nuclei of which the DNA as a salt-like compound with a basic protein is organized into chromosomes housed within a nucleus separated from the remainder of the cell by a distinct membrane.

EUPHENICS

This term is the coinage of Joshua Lederberg and stands for the attempted annulment of the effects of deleterious genes by improvement of the environment. An example is given in the entry PHENYLKETONURIA, where it is suggested that phenylalanine be reduced in the diet of children constitutionally unable to manufacture the enzyme phenylalanine hydroxylase that is needed to metabolize phenylalanine.

Contrary to popular belief, euphenics does not invariably have dysgenic consequences; some euphenic procedures actually bring about genetic improvement. An example is given under SICKLE-CELL ANEMIA.

EVIDENCE OF EVOLUTION

Various contemporary currents of thought have given rise in recent years to the impression, perhaps the hope, that the notion of evolution has somehow been discredited and that the doctrine of special creation has been reinstated and possibly even put on a scientific foundation. There is no substance in either view, though it is difficult not to sympathize with the layman's bewilderment upon learning that acceptance of the hypothesis of evolution does not rest—as he had assumed it must—upon the validity of so-called proofs of evolution, most of which are unconvincing or open to other interpretations, but rather upon evidence of a different and far weightier kind.

The best way to explain the difference is to consider why we believe that the world is "round" (spherical). When bright schoolchildren are first introduced to this notion, they are offered "proofs" of the roundness of the earth, which in retrospect seem pretty trifling; they are told that when a ship is sighted on the horizon they see first its mainmast or funnels, then its other superstructures; only later does the hull heave into view. This is most unconvincing—a mere trick of refraction, perhaps, such as one might expect at the level of the horizon. Another such nugget of evidence is that if three posts of equal height are stuck into the ground in a straight line, the middle post will look higher than the other two (very difficult here to arrange matters in such a way as to make such a "proof" technically acceptable).

Outside school our belief in the earth's being round does not depend on any one or two proofs. We accept the idea because the whole of navigation, aeronautics, geodesy, geometry, cartography, and chronometry rest upon it: it is not now possible to think rationally on any of these subjects without accepting the concept of the roundness of the earth. No scientist need convince himself of the truth of the matter by looking apprehensively—it might be thought—at photographs of the earth taken from the moon.

Essentially the same holds for the hypothesis of evolution; opponents of the idea, who are so often philosophically as well as scientifically illiterate, think it shifty and evasive to contend that acceptance of the hypothesis does *not* depend upon acceptance of certain proofs of evolution; but indeed it does not. We accept the notion of evolution because it alone makes sense of the pattern of similarities and differences among contemporary living organisms that is revealed by the study of comparative anatomy; of the phenomena revealed by or implied by von Baer's Law, such as the remarkable similarity between the embryos of human beings, birds, and reptiles, on the one hand, and on the other, the embryos of their reputed ancestors, such as fish. It makes sense also of the existence of what would otherwise be thought of as somewhat anomalous animals with intermediate characteristics, such as feathered reptiles, fish with lungs, or mammals that lay shelled eggs. It makes comprehensible too the evolutionary transformations we have witnessed in our own lifetimes, such as the spread of the melanic variants of moths in the countryside near industrial areas. Another example is the evolution in many hospitals of strains of staphylococci or streptococci that are resistant to the action of penicillin and other antibiotics. Finally, only the theory of evolution makes a convincing story of the fossil record. A man who believes that fossils are the remains of organisms inundated by Noah's flood can believe anything: no effort of credulity would be too much. Only evolution theory makes vestigial structures in contemporary organisms intelligible—as, for instance, the transient appearance during the development of higher vertebrates of a median (pineal) eye, such as we find in lampreys.

Opponents of evolution often think they have scored a point in saying that "evolution is only a hypothesis." Yet the word "hypothesis" is used in this illustration precisely because it is logically and philosophically the correct term with which to describe a proposition of this stature.* The word has no pejorative connotation whatsoever: it is a sign of lack of learning to suppose that it has.

There are still many uncertainties about the mechanism of evolution, and some at least about the exact course it took in the history of existing animals and plants. Nevertheless, the theory of evolution is scientifically acceptable, is indeed widely accepted: it is the product of an exercise of mind wholly different from that which gave rise to the moving, imaginative literature of the first chapter of Genesis.

EXOGENETIC HEREDITY

Exogenetic heredity is the form of heredity that is mediated not through the chromosomes but through other—generally speaking, "cultural"—means of information transfer. That which is propagated is knowledge, know-how, and various products of

---

* It is a provisionally believed-in proposition, in the logical position of the premise of a deductive argument from which deductive inferences may be drawn to see whether or not they correspond to real life.

mind such as books, laws, rules of behavior, and mechanical and other inventions. When human artifacts undergo a slow, systematic, secular change generation by generation, as they often do, then we may reasonably speak of an "evolution"—an exogenetic or, as Alfred J. Lotka once put it, "exosomatic" evolution, as opposed to evolution of the more familiar endogenetic or endosomatic kind.

Biologists who rate themselves straightforward, tough-minded, no-nonsense people are apt to dismiss the whole notion of exogenetic evolution as fanciful, woolly-minded, or (with a certain pejorative inflection) merely "philosophical." These wise folk should try to keep it in the forefront of their minds that the notion of exogenetic heredity was first propounded by a specially straightforward, tough-minded geneticist notorious for his lack of nonsensicality. Thomas Hunt Morgan (1866–1945) wrote on the matter thus:

While biologists have come to reject the theory of the inheritance of acquired characters by means of the germ-cells, nevertheless they recognize the fact that the human race has succeeded in another way in transmitting certain traits acquired in one generation to the next. There are, then, in man two processes of inheritance: one through the physical continuity of the germ-cells; and the other through the transmission of the experiences of one generation to the next by means of example and by spoken and written language. It is his ability to communicate with his fellows and train his offspring that has probably been the chief agency in the rapid social evolution of man. In the animal kingdom we find many cases in which the young are protected and cared for by their parents. Such beginnings furnish the background out of which has evolved the more complex relation of parents and offspring in the human race, where a prolonged period of childhood furnishes exceptional opportunities for the transmission of tradition and experience. (*The Scientific Basis of Evolution* [London: 1932], chap. 10)

This kind of heredity, and the form of evolutionary change that is mediated through it, are sometimes referred to as cultural heredity or evolution. It is an injudicious designation, for it gives rise to the impression that what is evolving is culture, whereas the terminology we have adopted makes it clear that we are speaking of a scheme of heredity and of evolution that is mediated *through* culture. Exogenetic evolution, while an arresting notion, is also an obvious one—something that most biologists have come upon or worked out for themselves without prompting from Morgan, Herbert Spencer, or any of the other advocates of the idea. Perhaps its very obviousness is what has made it unattractive as a subject of investigation.

Exogenetic evolution is that which has made man not merely a terrestrial creature, but also an aerial, marine, and submarine creature—which has brought it about that men may live at the equator or at the poles, in space, and one day, no doubt, elsewhere in the solar system. It is exogenetic evolution too which has brought it about that human sensibilities have been so far refined that by the use of one or another exosomatic instrument we may see objects many billions of times smaller than the smallest object that can be discerned by the naked eye and many millions of miles farther away than the most distant we can ordinarily see. Sophisticated instruments enable us to sense X rays, moreover, and discern the heat of a candle burning four miles away.

Microscopes, telescopes, Geiger counters, clothes, levers, wrenches, hammers, cutlery, and guns are among the exosomatic organs of mankind, and these evolved much as our own proper organs evolved. There are some amusing and deeply instructive parallels between the evolution of endosomatic and of exosomatic organs: both are gradual and both may embody vestigial characteristics, especially in clothing (one thinks of those long since functionless buttons at the ends of the sleeves of jackets worn by men and of the buttonholes in their lapels). Again, as they wear out or are superseded or discarded, exosomatic organs have to be produced again by manufacturing processes—and natural selection has some part to play here with respect to their economic fitness: only that which works and which people want enough to buy will be reproduced anew, and a fresh variant will soon become the prevailing type in the population. Red in tooth and claw, the marketplace has witnessed the complete dispossession by the ballpoint of the goose quill that so long served our great-grandparents.

Mutation plays in exosomatic evolution a role comparable to the one it plays in ordinary evolution: when John Boyd Dunlop (1840–1921) invented the pneumatic tire, it conferred such a high degree of economic fitness that it swept through the population of vehicles of all kinds so rapidly that a vehicle with solid tires has become a curiosity within our own lifetime.

These various parallels are not to be rated much higher than fun, but the differences between exosomatic and endosomatic evolution are profound: for one thing, exosomatic evolution is fully reversible and for all practical purposes endosomatic evolution is not. Again, as Morgan clearly foresaw, exosomatic evolution is in the Lamarckian style, for it embodies a learning process. Endogenetic heredity does not—and what is learned can become part of the heritage. The most compendious description of the content of exogenetic heredity is Karl Popper's "Third World," for obvious reasons now referred to as "World 3." (See Popper's *Objective Knowledge* [Oxford, 1975].) Where World 1 is the ordinary physical world and World 2 is the world of our conscious experience, World 3 comprises the products of the human mind—the world of memories, programs, rules and instructions, arguments, theories, the content of books, and so on. In the light of this description, essentially Popper's, we can see the force of Richard Dawkins' conception of the *meme* as a unit of exogenetic heredity.

The reaction of many biologists to learning about exogenetic heredity and evolution is often "So what?" There is no need to elaborate here upon the answer that it is through exogenetic heredity that human beings owe the greater part of their present fitness and their hope of becoming fitter still.

FERTILITY

In demography "fertility" means reproductive performance; reproductive potential or capability is referred to as "fecundity"—a point worth making, because biologists tend to use these words the other way around. By a natural extension of meaning, eggs are described as fertile or infertile according to whether they are or are not capable of developing into whatever they are the eggs of. Matters to do with reproduction are dealt with elsewhere, so the present discussion will limit itself to problems of fertility as they arise in demography.

During the 1930s the fertility of the population of the United States and the countries of northern Europe, continuing a trend that had begun at least fifty years before, sank to so low a level that doubts were expressed about the ability of these populations to maintain their numbers. The "twilight of parenthood" was widely spoken of, and the world's leading demographer was confident that the Western world was facing doom in spite of the fact that population numbers were not falling. Laymen in general were not at all alarmed, not realizing that the increase, or lack of decrease, in population numbers was due to the steady accumulation of people beyond reproductive age. Demographers themselves became very interested in devising measurements of the reproductive vitality of a population, of its ability to maintain its numbers with regard to the prevailing regimen of fertility and mortality.

One may ask of any member of the population, for example, what is the likelihood that he or she will be represented by an individual surviving to the same age in the next generation? If we ask this question of the female population only, we are in effect asking, what is the ratio of live female births in successive generations? The figure is Robert Kuczynski's "net reproduction ratio" (NRR). If the NRR is greater than one, the population is more than holding its own; but if it is less than one, the population is declining. Another such measure is Alfred J. Lotka's "true rate of natural increase," renamed by R. A. Fisher the "Malthusian parameter," which expresses the rate of continuous compound interest at which the population is changing in size with respect to the prevailing rates of mortality and fertility. If the value is positive, the population is holding its own; if negative, it is not.

The attempt to express the reproductive vitality of a population by a single scalar measure—as if we were taking its temperature—is doomed to failure: it exhibits the same wide-eyed innocence as the fervor that has prompted some psychologists to believe that intelligence can be measured by a single scalar figure such as the IQ.

For purposes of population projection demographers have now adopted a quite different method, in which *average completed family size* is the quantity considered most significant. This scheme of measurement is much easier to resolve into independent variables that have a real meaning in terms of the way people behave—in terms, for example, of marriage age, marriage rate, and pattern of family building.

FIGURES OF SPEECH

The familiars of the living world have greatly enriched everyday speech with metaphors and similes, not all of which are strictly fair to the organisms that gave rise to them. "To drink like a fish" is based on an obvious misunderstanding: a fish's apparent drinking is the swallowing movement by which it keeps up a respiratory current of oxygen-bearing water through the gills.

It is less clear why newts are reputed to be under the influence of liquor, as in the well-known English euphemism "as overtired as a newt." We may attribute it to the somewhat tipsy-looking side-to-side movement of the newt's relatively big head, produced by its sigmoid swimming movements.

"Alike as two peas" is surely disrespectful to the memory of Abbé Gregor Mendel, whose niche on the slopes of Parnassus was made possible by the *un*likeness of pea seeds and pea plants.

"As dead as mutton" leads irresistibly to the reflection that

Cambridge, England, is hardly less famous for making sausages than as a seat of learning, and that one of the last earthly acts of Thomas Strangeways, eponym of the famous tissue culture laboratory on the outskirts of Cambridge, was to culture surviving fibroblastic cells recovered from the inside or *endosarc* of a sausage, thereby showing that individual cells may outlive the organism of which they were at one time a part.

"As mad as a hatter" is sociological rather than biological: it refers to the mental impairment that was at one time brought about in hatters by toxic chemicals used in the dressing of fur. "As mad as a March hare" may perhaps justly impute madness to the hare's single-minded absorption in its mating exercises.

The recruitment of new figures of speech and similitudes into the English language is a continuous process, as vigorous today as ever before. When a biologist hears someone's table manners being compared unfavorably with those of a piranha, he or she can at the same time hear the little chink which signifies that biology has dropped another jewel into the treasure chest of the English tongue.

FITNESS

The importation into science, with special technical meanings, of words that have a long history of vernacular usage is a constant source of confusion and embarrassment. The word "fitness" is a case in point, for only the connotation of suitability or *adaptedness* in the vernacular usage persists in the technical meaning. The other connotation, which calls to mind middle-aged individuals jogging around the block or taking deep breaths at an open window, is not part of the technical meaning.

Fitness is the characteristic of organisms, of populations, or of genes that is maximized by the operation of natural selection. The quantity or property maximized is *net reproductive advantage,* the word "net" emphasizing that the measurement of fertility makes allowance for mortality (see "net reproduction ratio" under the entry FERTILITY).

It was this usage of the word "fitness" that prompted Herbert Spencer to refer to natural selection as the agency that brought about the "survival of the fittest." Intelligent schoolchildren are quick to spot the element of tautology in this description, which might be construed to mean the "survival of those that survive"; it is a shame, though, that those bright youngsters, far from fulfilling their early promise, often turn into adults who repeat this criticism as if it were original or in any real sense damaging. The notion of net reproductive advantage is explained and exemplified in the entry having to do with NATURAL SELECTION.

A wise biologist has pointed out that net reproductive advantage is not the only conceivable measure of fitness in the more general sense of adaptedness; the fitness of a mammoth's woolly coat might well be measured by its thermal conductivity (the lower the better), but it is still appropriate to say that unless a thick woolly coat had conferred Darwinian fitness upon its possessors—or, if we choose to put it that way, upon the genes that are responsible for its formation—then mammoths could not have evolved into possession of it. The recently introduced idea of *inclusive fitness* makes explicit allowance for the fact that the net reproductive advantage of a gene will necessarily include any advantage secured vicariously, in the sense in which parental care and certain types of family cooperation may provide for the propagation of an individual's genes. "Inclusive fitness," an authority has written, "may be defined as that property of an individual organism which will appear to be maximised when what is really being maximised is gene survival."

Because of its adoption into technical language from vernacular speech, the Darwinian conception of fitness is still the subject of grumbling criticisms. At least some of these result from a misunderstanding of the notion, in particular the belief that fitness is measured only in terms of fertility, with the implication that all genetic changes that provide for enlarged fertility must of necessity prevail.

FORCE OF MORTALITY

The force of mortality is the age-specific death rate. In a population not subject to senescence (deterioration with increasing age) the force of mortality will be constant, but in real life the characteristic pattern is one in which the force of mortality starts high in the first year or two of human life because of the special dangers associated with childbirth and extreme infancy. It then falls to a minimum at about the age of fourteen or fifteen, the actuarial prime of life, in which a human being is more likely to live an extra year or month or day or minute than at any other age. This is the epoch, too, at which a human being's reproductive value is highest. After this period of prime, the force of mortality ascends inexorably until our closing years, for we do indeed become more vulnerable as life goes on.

FORM AND MATHEMATICS

The application of mathematical methods to the description and analysis of biological form is an enterprise that has had a powerful appeal for quantitatively minded biologists. It is a subject upon which two extreme and apparently irreconcilable views have been held. Sir D'Arcy Wentworth Thompson, author of the famous essay *On Growth and Form* (Cambridge, 1917) wrote:

The study of form may be descriptive merely, or it may become analytical. We begin by describing the shape of an object in the simple words of common speech: we end by describing it in the precise language of mathematics; and the one method tends to follow the other in strict scientific order and historical continuity ... The mathematical definition of form has a quality of precision quite lacking in our earlier stage of mere description ... We are brought by means of it into touch with Galileo's aphorism (as old as Plato, as old as Pythagoras, as old perhaps as the wisdom of the Egyptians) that the book of nature is written in characters of geometry. (p. 719)

The antithetical view was admirably well put by David Meredith Seares Watson, one of the foremost paleontologists of the twentieth century in his Silliman Lectures, *Paleontology and the Modern Biology* (New Haven, 1951):

The existence of real, fundamental structural resemblance can only be established by using the intellectual processes which characterise that branch of zoology called morphology. Morphology is a form of logical thought remarkable in that it is not mathematical; indeed, its essential elements, being, as they are, qualities, are not susceptible of mathematical expression. (p. 3)

In the course of the next few paragraphs we hope to show that there is an element of truth in both viewpoints.

Consider first the mathematical representation of form. It follows from Cartesian principles that a system of surfaces in space can in principle always be represented by a system of functional relationships among three variables, x, y, and z, the relationships being conventionally expressed in the form $F(x, y, z) = 0$.

Thus the horns shown in the left-hand figure below may be represented algebraically by a function

$$(x + cy^2)^2 + (z - y^{2/3})^2 = a(k^2 - y^2)^3.$$

Similarly, a paraboloid with cusps (center below) and its more artistic representation (on the right) are spatial realizations of the formula

$$\left[ax + \left(z^2 - \frac{a^2}{4} \right) \right]^3 \left[ax - \left(z^2 - \frac{a^2}{4} \right) \right] + k^2 a^4 y^2 = 0.$$

With these examples in mind (all from the Mathematics Department of the Science Museum in London, by whose kind permission they are reproduced here), how can it validly be said that form is not susceptible to mathematical treatment? Watson was right, though, for in the examples we have chosen, there was no intention of trying to find a mathematical description of given forms such as we illustrate. On the contrary, the formulas came first and the figures are simply their geometric realization. Under the influence of the geometers Julius Plücker and Jacob Steiner, the spatial realization of complex algebraic relationships

became something of a cottage industry. It was a patient Irish lecturer in the old City and Guilds College in London who modeled the formulas quoted in the text: it would be the labor of a lifetime and require a battery of computers to describe mathematically the shape of even a quite simple organism—and the performance of such an exercise would be completely pointless and teach one nothing. After all, it is not very often that we need a minute description of a given form. More often, especially in biology, what we want to do is *compare* forms. This is an entirely feasible enterprise mathematically—made possible by the method of transformations which, with its ancillary notions such as variance and invariance, is capable of throwing light on such fundamental biological concepts as homology.

Consider the image of a lantern slide thrown on a rigid screen, and imagine that the screen is tilted this way or that from its normal position at right angles to the optical axis of the projector and the cone of light that issues from it. Tilting the screen would cut the cone of light at different angles, and the image projected upon the screen could be distorted in a way and to a degree that would depend on the direction and the degree of tilt; by this means we could generate a whole class of geometric figures: circle, ellipse, parabola, and hyperbola, and in an extreme case two intersecting straight lines—in fact, the *conic sections*. These are generated by "projective transformations," which are susceptible to an entirely precise mathematical definition even if we cannot define mathematically the form of the image that is projected (which might be of a landscape or a portrait or a whole animal). Change of shape can be mathematically defined, even if shape itself cannot.

Because it is independent of the subject to which it is applied, a transformation may be thought of as an operation in its own right regardless of its subject. The mathematical expression that defines the characteristics of and gives the rule for the operation is known as a *mapping function,* as explained in the entry on TRANSFORMA-TIONS.

With the projector model in mind, the distinction between the properties that are variant or invariant under transformation becomes self-evident. The property of being a circle or a square has no meaning in projective space, because under projection a circle becomes an ellipse or other conic section, and a square a rectangle; however, the property of being a conic section is clearly invariant under transformation. The truths embodied in the theorems of projective geometry are also invariant under projective transformation—in fact, projective geometry is so defined. Two other invariant properties may be mentioned: *linearity* is one, for that which is straight on the slide will be straight on the screen, no matter how we tilt it, and certain special ratios such as the cross ratio or anharmonic ratio are also invariant. *Parallelism,* however, is not. It was an important episode in the history of geometry when Felix Klein (1849–1925) defined a geometry as a study of properties of space that are invariant under a specified group of transformations. A "group" of operations in this sense is a set of operations such that the product or consequence of applying any two operations of the set successively is itself a member of the set.

Many morphological concepts which we grasp intuitively can be defined formally in terms of transformation theory. Let us ask, for example, what is the minimal affinity that must hold between two forms if they are to be comparable at all? This question has a precise answer: two forms are comparable when they are *homeomorphic,* that is, related to each other by continuous one-to-one mapping functions. Geometrically this means that the mapping function which defines the transformation and so assimilates the two forms must be such that each point in one is brought into correspondence with one and only one point in the other and that the function is continuous. An intuitive realization of these requirements would be the transformations brought about in a drawing on a sheet of rubber that is expanded or contracted or turned or twisted in any way we please, *provided the rubber is not torn.* Figures so related are said to be homeomorphic; thus the two forms immediately below are homeomorphic; so are the next

three, all metric relations being now lost. One might well wonder if *any* properties remain invariant under these highly permissive transformations. Some do, for certainly a basic affinity remains: the relationships of insideness and outsideness, and the relative placing of landmarks are the main invariants. So is Euler's Theorem, illustrated by the first set of homeomorphs: The number of enclosed spaces plus the number of meeting points equals the number of lines plus one. The homologies in these frames are easy to discern.

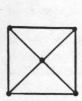

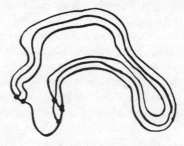

Contrast with these sets of homeomorphic figures the ones now illustrated. No process of continuous transformation could transform these into each other. They are nonhomeomorphic—of different topological rank. If these were organisms we could allot

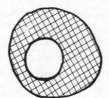
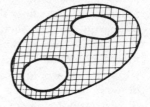

each one to a different taxon. Homeomorphy is that very basic similarity of ground plan which unites the members of a single phylum, an affinity which disregards differences that are not differences of ground plan.

In summary, we believe that although the mathematical treatment of form as such is usually not possible, the exact mathematical treatment of *change* of form very often is. We also believe that some of the ideas of abstract geometric theory can illuminate the fundamental biological concepts of morphology. Thus there is a great deal of truth in *both* of the antithetical opinions quoted at the start of this entry. For a much more sophisticated and versatile treatment of the mathematics of form that is still basically

Kleinian in approach, consult *The Fractal Geometry of Nature* by Benoit B. Mandelbrot (San Francisco, 1982).

FRAUDS

Frauds in science are conscious and purposeful misrepresentations of the truth for personal gain. We cannot distinguish them morally from ordinary crimes that have a scientific setting, such as plagiarism, theft of data or ideas, forging of laudatory testimonials by jobseekers, and so on.

Discovered scientific frauds are always considered especially shocking, and because only a minority can ever be discovered they are thought to cast doubt on the probity of the entire scientific profession ("tips of icebergs" are referred to by people for whom clichés are a habitual substitute for thought). Clearly, this is most unfair: who knows how many venal policemen there may be, or secret diabolists among those in holy orders, or college presidents secretly opposed to learning in all forms? But we think these are the exceptions, not the rule.

It is not surprising that so many discovered frauds in science are committed by practitioners of the life sciences. In the physical sciences everything is so much more cut and dried, and the subject matter so much less variable. If a scientist eager for self-advancement were to publish a fictitious statement about the physicochemical properties of cetyltrimethylammoniumbromide or of triiodothyronine, he would not long remain undiscovered; his claim would be found to flout some canon of physics or chemistry, and his work could be repeated and confuted without great exertion. No such comfortable words can be said of the notorious biological frauds: the Piltdown skull, for example, which purported to show that England was a cradle—not merely a throne—of mankind; and the deeply discreditable story of the white mouse made up with a felt-tip brush to create the illusion that it bore a living skin graft from an unrelated black mouse. The latter was an episode that caused immense trouble to workers in the same field and diminished the reputation of a laboratory head to whom the criminal owed nothing but gratitude. Then there was the famous case of a professor of psychology at a London university who manipulated figures relating to the IQs of twins in such a way as to support his belief that differences of intelligence were due in far greater measure than was usually supposed to heredity as opposed to upbringing. His scientific crimes were eventually uncovered by a skillful investigative journalist with training in genetics. Perhaps the most notorious biological forgery was that associated with the name of Paul Kammerer, who professed to have demonstrated Lamarckian inheritance in midwife toads.

No commonplace explanation can account for all these frauds. The motive underlying at least three of them was a passionate belief in the truth and wider-than-scientific importance of a right-seeming doctrine that lay so far outside the accepted canon that it was felt to be the subject of unfair discrimination by orthodox scientists. Indeed, belief in the Lamarckian style of inheritance burns in its devotees with the kind of intensity that literary people regard as almost a guarantee of authenticity. We have Coleridge's assurance, as we had Plato's, that there is something divine about creative inspiration; some such thought may console the tricksters when they find that Nature is dragging her feet.

Scientists have been pronounced simple-minded and gullible for being taken in by these frauds. There have even been hints that they close ranks and do their best to hush up anything that might diminish their supposedly sacerdotal reputation. This is a very superficial view. It gives no weight to the important truth that science, like banking, politics, marriage, and indeed most human relationships, can only be conducted on a basis of confidence. We cannot carry on our life forever on our guard. The field naturalist who shyly brings a fossil to the museum does not expect to be greeted with "Another forgery, I suppose!" nor does the scientist who proudly submits his latest paper to his department head expect to hear "I'm glad to have your paper; you know what a devoted reader of science fiction I am." Science could no more prosper under such conditions than could a marriage in which a husband late from work was invariably met by the words: "Fornicating again, I'll bet!"

GENETICISM

A deliberately pejorative neologism, "geneticism" stands for the belief that in the shaping of man and of mankind generally, inheritance and genetic endowment are not merely important, they are all-important. For its practitioners genetic explanations must be sought not only for character differences of the kind studied by Mendel but for the rise and fall of nations, the stratification of society, the existence of national differences in pronunciation (such as the ability or inability to pronounce *theta* in a way that differentiates it from *zeta*), and—most assuredly—differences in intelligence. Indeed, in complete disregard of the considerations put forward in our entry on NATURE AND NURTURE intelligence has been pronounced to be some 80 percent heritable (that is, genetically prescribed).

One of the gravest and most widespread aberrations of geneticism is embodied in the belief that if any characteristic is enjoyed by *all* individuals of the community, it must be genetically underwritten. Thus, if it should turn out that a certain basic linguistic form such as the Aristotelian subject/predicate form is an element of all the languages of the world, then its usage must be genetically programmed. (Some of Noam Chomsky's writings are not guiltless of this assumption, which is also a disfigurement of sociobiology as it steers its precarious course between the twin perils of geneticism and historicism.) It may be well to repeat in this context the reason why this supreme canon of geneticism is not satisfactory: if any trait is to be judged "inborn" or genetically programmed, then there must be some people who lack it. The ability to taste phenylthiocarbamide, for instance, is known to be genetically programmed because there are those who lack it.

The practice of geneticism leads to a systematic depreciation of, indeed an active enmity toward, the idea of cultural heredity as described in our entry EXOGENETIC HEREDITY. The importance and character of cultural heredity were first recognized by Thomas Hunt Morgan, second only to Mendel in importance in the history of genetics. Morgan's delineation of what is now sometimes called psychosocial evolution was scathingly dismissed by C. D. Darlington, then professor of botany at the University of Oxford and a notorious practitioner of geneticism. See his review in *Nature* 278 (1979):786–787 of Morgan's *Scientific Basis of Evolution* (New York, 1932). Morgan's reputation has, however, outlived Darlington's, and the notion of exogenetic heredity will doubtless outlive them both.

GENETIC AND EMBRYOLOGIC TERMS

Biologists in general, and geneticists and embryologists in particular, tend to be meticulous in their use of technical terms and shun such slovenly turns of speech as "the gene for albinism" or "the gene for tallness." Careless usage tends to obfuscate the nature of the interaction between the gene and the environment in which it expresses itself. We therefore include a set of terms used in genetics and embryology with the explanations we deem appropriate.

ALLANTOIS The excretion of waste matter other than gases is quite a problem for vertebrate animals that lay shelled eggs. Their repository for fluid or semifluid waste is an embryonic organ, the allantois, that forms as an outpushing from the hind end of the embryonic gut, which abuts into either the yolk sac or the cavity of the amnion. From the way in which it is formed, the allantois of necessity has an outer lining of richly vascular connective tissue. This tissue plays a most important part in formation of the placenta of the higher mammals, in which an allantois is also present and where development is much like that of a reptile in its early stages.

ALLELE Gregor Mendel's pioneering genetic researches were made very much easier than they otherwise would have been by the combination of luck and judgment that led him to study the inheritance of differences between pairs of contrasted characters, such as tallness vs. shortness in pea plants and their possession of yellow or green seeds, which might themselves be either round or wrinkled. Thus the conception took root that genetic factors were essentially present in pairs responsible for a pair of contrasted characters.

This state of affairs is by no means typical. Genetic factors customarily form not pairs but sets—of which, of course, not more than two can be represented in any one organism at any one time. Such sets of characters are referred to as allelomorphic (Greek *allos* = of another sort or kind). Mendel studied the special case of binary allelism, but multiple allelism is the more general case: thus the factors bringing about the color dilutions of which the extreme form is albinism form an allelic system.

AMNIOCENTESIS This process involves the withdrawal of a few drops of amniotic fluid during pregnancy to provide fetal cells that can be cultivated in order to ascertain the structural normality (or lack of it) of the makeup of the chromosomes. By such means an abnormality such as Down's syndrome can be identified before birth. This diagnosis can then be made the basis of a decision about the future of the embryo. The procedure, often adopted for the fetuses of older mothers, brings with it the unexpected bonus of informing the parents of the sex of their child, thereby enabling friends and relatives to decide upon a name for it.

AMNION Although vertebrate animals are rightly said to have evolved from aquatic into terrestrial or aerial animals, none is wholly emancipated, for all vertebrate animals develop in what is actually or virtually an aquatic environment. The embryonic organ that makes this feat possible is the amnion, a bag-like structure formed from the outermost of the various envelopes that surround the developing embryo. It is lined with connective tissue, in which run blood vessels connected to the embryo's heart. These various envelopes enclose a space, the amniotic cavity, filled with a fluid that is in all respects—osmotic, ionic, hydrostatic—a suitable environment for embryonic development. The embryo's mouth is open to the amniotic fluid, which accordingly passes freely into the fetal gut. All vertebrate embryos are thus aquatic organisms.

Vertebrate animals possessing an amnion are referred to as amniotes: reptiles, birds, and mammals are so described and the description is taxonomically more sound than "tetrapod," a taxon into which birds fit uneasily.

ASEXUAL REPRODUCTION Reproduction that is not mediated through gametes, but through reproductive cells such as spores or by budding from the parental stock, is known as asexual. It leads to the formation of clones of organisms of the same genetic makeup as their progenitor, so that there is no increase of genetic diversity. It is quite common among sessile animals and in plants—hence the term "vegetative reproduction," often used for asexual reproduction in animals. It is closely related to the faculty of regeneration and naturally lends itself to the formation of colonies attached to one another, as in corals and in polyps. It is quite distinct from parthenogenesis, a variant of sexual reproduction. Twinning of the kind described as identical in our definition of TWINS represents the only kind of asexual reproduction found in vertebrate animals.

ASSORTATIVE MATING   A technical term used especially by students of human heredity to refer to the mating of like with like with respect to some specified natural or nurtural characteristic. Thus any tendency of men and women of similar stature, intelligence, blood group, religion, or socioeconomic class to mate with each other would qualify as assortative. Mating predilections, whether assortative or nonassortative, clearly have an important influence on the genetic makeup of human populations: orthodox Jews, for instance, would lose any racial distinctiveness they still possess were it not that they incline to marry others of the same faith.

CONGENITAL TRAIT   Taken literally, the word "congenital" refers only to that with which a human being is born. Common usage, however, tends to restrict the term to departures from normality. No one is likely to be gratified to learn that he has congenital ears. There is a wide variety of congenital aberrations, some of which, for example ANENCEPHALY and DOWN'S SYNDROME, are the subject of separate entries.

CROSSING OVER   Just as genetic linkage modifies the law of independent assortment, so the phenomenon of crossing over modifies the principle of linkage in the direction of independent assortment. Linkage groups do not typically maintain their integrity, for the matching chromosomes of a pair tend to exchange material and so recombine the linked genetic factors of which they are the vehicles, a process that enormously enriches genetic variance.

DOMINANCE AND RECESSIVENESS   If the somatic effect of an allele only shows up when like alleles have been inherited from both parents, the genetic trait for which that allele is responsible is said to be *recessive* in expression; shortness in pea plants is a case in point, one investigated by Mendel. Phenylketonuria in human beings is described as a recessive disease, because the factor responsible for it must be inherited from both parents if the disease is to become apparent. A gene which manifests itself even when inherited from only one parent is said to be *dominant* in expression; thus the heterozygous progeny of a cross between tall and short pea plants were found by Mendel to be tall. Some human diseases, among them Huntington's chorea, represent the outward expression of a dominant gene—that is, one that need have been inherited from one parent only. Obviously, dominance and recessiveness do not represent moral valuations: they are simply statements about the mode of expression of genetic factors. Both states can be modified by other genetic factors. Indeed, genes which are halfway between dominant and recessive—which have some expression in the heterozygous state—may evolve toward complete dominance or complete recessiveness of expression.

Although according to the book it should not be possible to distinguish a homozygous dominant from the corresponding heterozygote, a recessive factor present in the heterozygous state may sometimes exert an effect strong enough to make the heterozygote distinguishable. This applies, for example, to the heterozygous carriers of some damaging recessive diseases: the heterozygote "carrier" of the recessive gene responsible for phenylketonuria may have a less than normal phenylalanine tolerance that can be detected by the appropriate clinical tests—a consideration of some importance in the prevention of phenylketonuria by the discouragement of matings between carriers.

The most elementary genetic algebra shows that when two heterozygotes of a genetic factor are mated, half their offspring will be heterozygous, as their parents were, and the other half will be homozygous for the two genes that entered into the hybrid, the formulation being $(Tt \times Tt) = TT + 2Tt + tt$.

ENTELECHY Aristotle's "entelecheia" was that principle or spirit which informs a structure and completes it in some essential respect, making actual that which had previously been merely potential.

In his Gifford Lectures of 1908 the renowned German experimental embryologist Hans Driesch (1867–1941) centered his doctrine of vitalism on a variant of the notion of entelechy that retains the sense and thrust of the Aristotelian usage: the organism is something more than its substance and form; it is *alive* by reason of its possession of the vital principle entelechy. Our entry on VITALISM discusses this conception. The sense of completion essential to the Aristotelian usage can also be found in Paul Tillich's (1886–1965) declaration that *courage* is the entelechy which completes a man.

EPIGENESIS In their older usages "epigenesis" and "preformation" were opposites; where preformation implied the unfolding of a structure already present, epigenesis referred to the acquirement of structure and differentiation in a system ostensibly structureless, as the egg was thought to be. Now that the egg is known to be highly structured at a molecular level, the antithesis no longer makes sense. In the modern usage "epigenesis" stands for all the processes that go into implementation of the genetic instructions contained within the fertilized egg. "Genetics proposes: epigenetics disposes."

GAMETE   A gamete is a reproductive cell, distinguished genetically by possessing half the number of chromosomes present in the ordinary somatic cells of the body. The gametes are egg cells (ova) in females and sperm in males.

GENE   The word "gene" was considered by the man who first used it, Wilhelm Ludvig Johannsen, and by Thomas Hunt Morgan (1866–1945), the famous American geneticist, as a physical embodiment of Mendel's determinant "factors" in heredity. It was nevertheless treated as a genetic entity, genetically defined. A gene was that which was responsible for a unit character-difference—in the Mendelian canon, for example, for tallness as opposed to shortness in edible pea plants, or green as opposed to yellow seeds. Thus explained, a gene appears as a unit of *segregation;* in other systems it might be defined as a unit of *mutation* or of *crossing over.*

This purely genetic conception of the gene remained adequate for the interpretation of breeding experiments for many years. The absence of a physical definition was not, and was not thought of as, an impediment to the growth of genetics. It was not indeed until the correct conception of the nature of genes was happened upon that a physical definition became possible. Genes have variously been thought of as templates, primers, models, or representative samples; as huge macromolecules; as bead-like *micellae* somehow linearly strung; and so on.

Genes are not adequately described by any of these terms, for genes are *messages*—a conception first propounded for chromosomes by Erwin Schrödinger and for genes by Hans Kalmus in the *Journal of Heredity* 1 (1950):19. Looking for a signal is an enterprise very different from looking for something of which one might make a microscopic preparation or an extract or a solution. As with many other questions, the answers came from molecular biology.

GENE FREQUENCY   The percentage or proportional representation of a gene in the population is the gene frequency, counting its homozygous representation as two and its heterozygous representation as one.

Thus if A and a designate a pair of contrasted Mendelian genes, in a population in which the genotypes occur in the proportion AA, AA, Aa, Aa, Aa the frequency of the A is 0.7 and of its allele necessarily $(1 - 0.7)$ = 0.3. Given random mating, these gene frequencies will be indefinitely conserved under the Mendelian regimen of inheritance. The Hardy-Weinberg Theorem makes it possible to translate statements about gene frequencies into statements about the frequencies of genotypes.

GENETIC CODE   In the entries on MOLECULAR BIOLOGY and NUCLEIC ACIDS we make clear that genetic information is conveyed by the linear ordering of nucleotides down the length of the molecule of nucleic acid. Taken in isolation, this statement is no more informative than to say that in old-fashioned Morse telegraphy a message is communicated by the order in which short and long electrical impulses or light flashes succeed one another. In fact, the genetic code is known as precisely as the Morse rendering of "M'aidez" (= Mayday), the international signal of distress.

In the genetic code particular amino acids are specified by a triplet of nucleotides; as there are four different nucleotides, sixty-four different triplets are possible. There is therefore some synonymy in the coding system, even allowing for special signals such as "message begins" or "message ends."

The genetic code and the main lines of its transcription and translation into the distinctive structures of various proteins are the same in plants, animals, and bacteria. Evidently, then, it is a very ancient inheritance of the world of living things.

GENETIC DRIFT   Most theorems in population genetics assume as a baseline for the norm that the members of an interfertile population mate at random, so that every member has an equal chance of being counted among the ancestors of future generations. The Hardy-Weinberg Theorem makes it clear that when mating is random and the population is equally divided by sex, the gene frequency remains constant and so also do the proportions of the various genotypes. We may, however, envisage two kinds of divergence from this state of affairs. The first is when the departure from randomness is systematic and leads to an increase in the proportions of one gene or genotype as a result of natural selection or of the repeated reintroduction of a gene into the population as a result of recurrent mutation. The second kind of departure is random; it occurs when the vagaries of random sampling create a situation where one gene is represented in the filial generation more often or less often than one would expect in a strictly random mating. If chance has it that such a departure from randomness occurs in the same direction in several successive generations, then the gene frequency in the population may change through the operation of chance alone. This phenomenon, known as genetic drift, has been thought to be very important in small populations. In large populations the chances of being counted and of not being counted among the ancestry of future generations tend to equalize.

There is now not much doubt—or much reason to doubt—that drift occurs, but the magnitude of the effect and the scale of its contribution to evolutionary change are still in question. The notion of drift was vigorously opposed by R. A. Fisher, who held the lowest possible opinion of Sewall Wright (mainly, we believe, because of jealousy). Wright was a principal pioneer of population genetics. His English contemporaries were in the main a rancorous and disputatious lot—not big enough, most of them, to allot credit elsewhere than to themselves. Mendel's theory, Fisher declared, could have been propounded from first principles, and in any case the results he professed to have secured were in a statistical sense too good to be true.

GENETIC SYSTEM   The genetic system of a species is the whole of that which provides for the transmission of genetic information from one generation to the next. A most important element is the mating system, whether it involves inbreeding or outbreeding and whether or not it embodies any element of assortative mating. The pattern of chromosomal behavior is of course another element in the genetic system—for example, the frequency of crossing over.

GENETIC VARIANCE   The moiety of the variance of a population that arises from genetic differences between its members is known as the genetic variance. If the measurable properties of the members of a population are expressed in terms of deviations, positive or negative, from their mean value, then the variance of the measured characteristic, necessarily a positive quantity, is the mean value of the squares of the deviations from the mean. In Mendelian heredity there is no tendency for genetic variance to decline from generation to generation, as it would if blending inheritance were the rule. This is the theorem of the utmost importance in the Mendelian theory of the evolutionary process.

GENOME   A genome is the genetic apparatus of a species considered as a whole and as characteristic of it.

GENOTYPE AND PHENOTYPE   An organism's phenotype is its overt and manifest character makeup, whereas its genotype is its genetic composition in terms of Mendelian factors or genes or whatever term may be used for the material genetic determinants. The distinction is of crucial importance in genetics and practical breeding. Thus an organism that is heterozygous for a dominant gene may have the same phenotype as the homozygous dominant, but breeding experiments soon show that their genotypes are entirely distinct.

HARDY-WEINBERG THEOREM The population geneticists' practice of thinking in terms of gene frequencies has struck many naturalists as offensively abstract, but the algorithm embodied in the Hardy-Weinberg Theorem makes it possible to translate statements about gene frequencies into concrete statements about the frequency of occurrence of particular genotypes.

Consider a pair of contrasted Mendelian genes A and a present in a population in the frequencies $p = 0.7$ and $q = (1 - p) = 0.3$ respectively. The Hardy-Weinberg Theorem states that in a random mating population, equally divided as to sex, the three possible genotypes AA, Aa, aa will be represented in the population in the frequencies $p^2, 2pq, q^2$ respectively. Then the genotype frequencies will be AA, 0.49; Aa, 0.42; aa, 0.09. These frequencies remain constant from generation to generation, except insofar as some impressed "force" such as natural selection is brought to bear on the population—a finding of immense importance for evolutionary theory.

HETEROZYGOSITY AND HOMOZYGOSITY When two like alleles are inherited from an organism's two parents, the organism is said to be homozygous with respect to that allele. Thus, to choose an example of Mendel's, if T and t stand for the genetic factors determining the characteristics of tallness and shortness respectively in the edible pea plant, an organism with the factor makeup TT is said to be homozygous for T, and if its genetic makeup should chance to be tt then homozygous for t. The makeup Tt—resulting from the union of gametes bearing unlike factors—is said to be heterozygous or hybrid with respect to those alleles.

HYBRID VIGOR The accession of physical vigor (especially fertility, longevity, and resistance to disease) that is found in the progeny of crosses between inbred or partially inbred organisms is known as hybrid vigor. Explanations of the phenomenon turn upon the restitution of a predominantly heterozygous state. It thus applies with special force when two inbred and predominantly homozygous lines are crossed, but the concept has no application to crosses between members of different human populations, each of which is likely to be adapted to its own environment and in any event predominantly heterozygous. The special mythology associated with "purity" of breeding and hybridity by the lay public and among people with a predilection for eugenics is a rare muddle with no solid scientific foundation—as in the belief fostered by boys' literature that "pure breeding" is an assurance of courage, fortitude, and all manly virtues, whereas hybrids or "half-castes" may be confidently assumed to be cowardly and treacherous.

INBREEDING AND INBRED LINE   The mating system in which a lineage is perpetuated by matings between siblings—that is, between brothers and sisters, or between parents and their offspring—is known as inbreeding. The most extreme form occurs in those plants in which self-fertilization is possible (in most hermaphrodite animals it is not). The effect of inbreeding is to increase homozygosity, leading eventually, perhaps, to the establishment of a pure line. Inbreeding may have the effect of bringing together somewhat deleterious recessive genes in the homozygous form in which alone they can exercise actions normally masked by wild-type dominant genes. Thus inbreeding leads, as a general rule, to some degree of genetic deterioration—an immemorial empirical observation of stockbreeders. A degree of inbreeding (as in cross-cousin marriages) takes place in some human communities, and different religions vary in their permissiveness with regard to the amount of inbreeding allowed.

A lineage is inbred when the members are for practical purposes genetically uniform and as homozygous in their genetic makeup as inbreeding can make them. The progeny of a cross between two such lines, while also genetically uniform, are heterozygous with respect to every gene by which the two inbred parental lines differ.

Inbreeding is a procedure fraught with difficulties, leading as often as not to the demise of the inbred line through the accumulation in it of too many harmful recessive genes, no longer masked by wild-type dominant genes. Fertility, in particular, is often so severely impaired that it is simply not feasible to continue to propagate the inbred line.

INDEPENDENT ASSORTMENT   A gloss on the law of segregation, independent assortment affirms in microscopic terms that when chromosomes of matching pairs separate at gamete formation, they enter the several gametes independently of each other in random allocations. The same is true of Mendel's hereditary "factors" when they are born on different chromosomes.

LARVA   Any embryonic form that represents a well-defined station in the course of development may be referred to as a larva. Thus the terms *blastula, gastrula,* and *neurula* are used in entries having to do with the development of chordate animals. All are commonly referred to as larvae, although none is an independent free-living organism. The word is most commonly applied, however, to some independent free-living organism that differs more or less radically from the adult and is transformed into the adult by a proportionately radical metamorphosis, the most familiar examples being the caterpillars, grubs, and maggots that are so often the larval stages of insects. Many larvae—especially those of sessile animals and of parasites—serve a distributive function.

Larvae as such do not reproduce (but see the entry on NEOTENY) and are produced anew only through mediation of the adult. In some organisms, indeed, a larva may in some nebulous sense be as "important" as the adult—and an object of at least equal zoological interest. Such a judgment might be held to apply, for example, to the mayfly larva, for the ephemeral adult mayfly lives only a few hours and is in effect a reproductive organ of the larva.

It is a corollary of von Baer's Law that the larvae of related forms tend to resemble one another much more closely than the respective adults into which they develop; in a number of large groups characteristic larvae often betray the affinities of their respective adult forms (see, for example, our entry on BARNACLES). Crustaceans typically have a nauplius larva, mollusks a trochophore larva, and sea urchins and their relatives a pluteus larva. The most famous larva of all is, of course, the tadpole of frogs and toads, the metamorphosis of which (involving as it does extensive retrogressive changes, especially of the tail) is one of the minor miracles of nature.

The ascidian tadpole is the mobile larva of sea squirts and their kin, so called because of its relatively large head and relatively long wriggly tail. It is widely thought of in the form which by neoteny became the ancestor of higher chordates.

Larvae, it should be added, are subject to selection pressures peculiar to themselves and can undergo quite a high degree of independent evolution of specializations that often have to do with distribution. An example is the long sticky thread attached to the so-called glochidium larva of the swan mussel.

LINKAGE   Formally speaking, linkage is a modification of the law of independent assortment. The segregation of Mendelian factors and their random and independent entry into different gametes occurs only when the factors under investigation are carried on different chromosomes, a simpler situation than that which obtains normally. For chromosomes are the vectors of more than one genetic factor and such factors tend to be inherited en bloc in so-called linkage groups, of which the microscopic equivalent is a chromosome.

MUTATION Mutation is the inception of inheritable variation, whether in the germ cells (germinal mutation) or in any ordinary body cell that is capable of division (somatic mutation), thus giving rise to a mutant clone. The physical basis of mutation is a structural change in one or more of the nucleotides that encode the genetic instructions conveyed by DNA, or it may take the form of a structural change in a chromosome. The changed form to which a mutation gives rise is known as a mutant—for instance, a mutant gene or, by extension, a mutant organism.

A mutant gene is an allele of the wild-type gene from which it represents a departure. The wild-type gene is sometimes for convenience signified by the symbol +. Mutant genes breed true, of course, but may undergo a further mutation either to a new allele or to the original + form (back mutation).

Mutations occur "spontaneously," that is, from unascertained causes at a frequency of the order of one in tens of thousands of gametes, this frequency being increased by ionizing radiations such as X rays and gamma rays and by a variety of chemical substances, notably by alkylating agencies. All substances that react with or upon DNA tend to raise the frequency of mutation also.

Such agents are nonspecific in that they do not cause one mutation rather than another. The randomness of mutations is a source of considerable disquiet to amateurs of biology, yet it is fundamental to an understanding of evolution theory to realize that—as mentioned in the entry on LAMARCKISM—"mutations do not arise in response to an organism's needs nor do they, except by accident, gratify them." It is not quite fair to say that mutations are the "raw material of evolutionary change," for that sounds as if evolutionary innovations had to await the chance occurrence of a suitable mutant which, if favorable, would then be received into the genetic establishment. A more exact formulation would state that mutant genes provide for genetic diversity and that, stirred into the genome by sexual fusion, crossing over combined with ordinary Mendelian segregation yields a vast spectrum of genetic variation that is the subject of selective discrimination.

It is not a defining characteristic of mutations that mutants are deleterious, but they often are—and not surprisingly, considering that they represent a disturbance in the existing genetic equilibrium; nor is it a defining characteristic of mutants that they are recessive in expression, although they often are. According to R. A. Fisher's original theory, mutant genes are, generally speaking, "completely dominant." That is, they have some expression in the heterozygous state, whereupon they may evolve into dominance or recessiveness of expression according to whether or not their effects are favorable.

ONTOGENY AND PHYLOGENY An organism's ontogenesis is, literally, its coming into being, and its ontogeny is its development considered abstractly and as a whole. Its phylogeny, by contrast, is its coming into being in the course of evolutionary descent and thus may be taken to stand for its evolutionary history considered abstractly and as a whole. The Law of Recapitulation is sometimes compendiously expressed in the form "Ontogeny recapitulates phylogeny."

OOCYTE The egg cell or female gamete, known as the oocyte, has only the haploid number of chromosomes. It contains only one chromosome from each of the pairs characteristic of the female's tissues (a number restored to the regular, or diploid, condition by union with a spermatozoon itself containing the haploid number of chromosomes).

The size and structure of oocytes, varying enormously from one organism to another, depend on the amount of yolk the egg cell carries and on the number and nature of the protective envelopes around the egg. Some eggs are almost yolk free, and the whole substance of the egg after subdivision into cells participates in the development of the embryo; but the eggs of amphibians, reptiles, and notably birds carry yolk which in the course of development is relegated to an extraembryonic position, until it is enveloped by cells of the embryonic gut and thus can be used up during development. In heavily yolked eggs only a polar cap of the egg cell, the germinal disk, participates in the formation of the embryo. The egg of the ostrich must be the largest of all cells.

Unlike spermatozoa the oocytes of vertebrate animals are not replenished. Generally speaking, there is a finite allocation of oocytes at birth and this is used up throughout life. Under these circumstances any arrangements will be selectively advantageous that increase to a maximum the likelihood of fertilization of the oocyte, then safeguard it during the first few days of life. There is a premium too on prompt fertilization of an oocyte that has been shed. Emil Witschi and others have shown that the fertilization of overripe eggs may lead to a variety of developmental abnormalities. It is the evolution of sexual cycles that has maximized the chances of fertilization and embryonic survival.

There is another sense in which "young" eggs are preferable to older ones: in view of the formation of a complement of oocytes at birth, it is clear that the oocytes shed by older mothers have been longer at risk of exposure to all influences inimical to eggs—among them ionizing radiations and toxic chemicals, from the influence of which the oocyte may not be wholly sequestered. There is ample evidence that chromosomal abnormalities, especially those that give rise to Down's syndrome, are more common in the children of older mothers. Ordinary (fraternal) twinning also is more frequent among the offspring of older mothers than among those of younger mothers.

OUTBREEDING   The state of affairs characteristic of most free-living organisms in which mating occurs between organisms that are not sibs or blood relations, outbreeding is often enforced by genetic or developmental mechanisms in hermaphrodites and sessile organisms which might otherwise be especially prone to inbreeding.

PARABIOSIS   At one time a popular experimental stratagem, parabiosis is an inelegant and generally speaking uninformative procedure in which two animals are joined surgically in the hope that their blood circulations will fuse and become confluent. Our newer knowledge of transplantation has made it clear that if the partners in parabiosis are genetically different, no effective vascular union can be achieved, for each partner then is an allograft upon the other. Almost anything that can be done by parabiosis can be done by simpler and more effective methods involving the use of cellular transplantation.

PLACENTA   The higher mammals, accordingly described as placental, are distinguished by the possession of an embryonic organ—the placenta—that is the medium of nearly all physiological transactions between the mother and the fetus, providing for both respiration and nutrition. The placenta is formed by the apposition of vascular embryonic membranes to the wall of the uterus, the apposed surfaces being multiply convoluted in such a way as to greatly increase the area of the frontier between mother and fetus.

Placentas are, as it were, opportunistically formed in members of all classes of vertebrates except birds. Considered as an abstract point of design, a placenta might be formed by one or the other of the two extremely vascular organs that are attached to, are part of, but lie anatomically outside the embryo, namely the *allantois* and the *yolk sac.* Placentas are of both kinds, and both may be found among the marsupials: the marsupial cat, *Dasyurus,* has a yolk sac placenta, and *Perameles,* the bandicoot, has an allantoic placenta.

The higher mammals have allantoic placentas, and at one time it was taken for granted that the allantoic placenta was the medium for *all* physiological transactions between mother and fetus. It has become increasingly apparent that the yolk sac is also an important pathway, especially for maternal antibodies entering the fetus. In kangaroos in particular, instead of a placental arrangement, a nutritive secretion of the uterus is taken up directly into the yolk sac.

From the way in which it is formed, the placenta must be partly maternal and partly fetal in origin; the fetal element that is most closely apposed to the tissues of the mother is not richly endowed with those antigens that in organ grafts arouse a rejection reaction. For this and other reasons a mammalian fetus is not treated as a foreign or nonself body and is not immunologically rejected, though there are certain circumstances in which embryonic ingredients can immunize the mother (see our entry on RHESUS FACTOR).

POLYMORPHISM   Polymorphism is the stable subdivision of an intrabreeding population into different genetic types which tend to maintain their proportions from generation to generation and which are such that even the least frequent variant occurs too frequently to be accounted for merely by the pressure of current mutation. Familiar examples are the subdivision of human beings into the blood groups A, B, AB, and O or into the types broadly described as Rhesus positive and Rhesus negative. Less familiar is the subdivision of populations of human beings and of mice into different tissue transplantation groups, those designated HLA and H2 respectively. Other polymorphisms are the ability and inability to taste the compound phenylthiourea (phenylthiocarbamide), which to some people is tasteless and to others—so we are told— very bitter. Another and theoretically very important example of polymorphism is the subdivision of the populations in West Africa and the Mediterranean basin into people who have either normal hemoglobin (A) or the abnormal recessive variant (S), which manifests itself as sickle-cell trait in heterozygotes or as sickle-cell anemia in SS homozygotes.

The anomalously high frequency of what are sometimes extremely deleterious genes and the stability of polymorphism from generation to generation point to a genetic enforcement, for which there are several possible mechanisms. The variants, for example, may be mutually dependent for their own perpetuation. Such is the case with sexual dimorphism, which is genetically enforced.

The more usual state of affairs, however, is where polymorphism is inevitably kept in being because a heterozygote enjoys a selective advantage over either of the corresponding homozygous forms (see SICKLE-CELL ANEMIA). If the heterozygotes predominate, their progeny will be homozygotes and heterozygotes in equal numbers, a self-perpetuating state of affairs. Selective advantage is by no means always apparent and sometimes is more the product of wishful thinking than of any adequately demonstrated net reproductive advantage.

Polymorphism, at any rate, is not a rarity or a curiosity but part of the texture of all normal populations. Polymorphism of genes coding for structural macromolecules is found almost wherever it is looked for, a circumstance that casts doubt on the orthodox belief that heterozygous advantage is *always* the cause of polymorphism.

POPULATION GENETICS There was a genuine theoretical revolution in biology when the population-genetic conception of evolution supplanted the genealogical or dynastic conception. The latter had been difficult to overthrow because of the ingrained habit of envisaging the course of evolution in terms of a family tree ("Fishes begat amphibia, amphibia begat reptiles, and reptiles in their turn begat on the one hand birds and on the other hand mammals ..."). Population genetics changed all this. That which evolves is a *population*, not a pedigree—and that which results from evolution is not *a* different organism or *a* new "type," but rather a new population of which the members have a spectrum of genetic makeups that provides for superior adaptation to whatever the selective forces may have been. This way of looking at things is entirely different from that figured in the conventional genealogical tables of evolutionary descent and is discussed further in the entry on NEO-DARWINISM.

SEGREGATION Expressed in microscopic rather than in purely genetic terms, Mendel's Law of Segregation states that in the formation of gametes the chromosomes of a matching pair part from each other and enter separate gametes. In genetic terms, the factors responsible for the determination of pairs of contrasted characters in Mendel's classic experiments (round vs. wrinkled or green vs. yellow seeds; tall vs. short plants) enter into different gametes. The segregation of genetic factors in gamete formation, followed by the random reunion of gametes at fertilization, makes possible the genetic recombination brought about by sexual reproduction.

SEX DETERMINATION Sometimes taken to refer to the ascertainment of sex, sometimes to the prior choice of sex, and sometimes to the process that causes an animal to belong to one sex rather than the other, "sex determination" is a very ambiguous phrase. We shall begin with the third sense.

In vertebrates the primary differentiation of sex is brought about by the segregation of a supernumerary pair of chromosomes known as the sex chromosomes in distinction to the others, the "autosomes," which perform ordinary somatic functions. In mammals generally, and certainly in human beings, the male sex is distinguished by possession of an unequal pair of sex chromosomes, designated X and Y. The male genotype with respect to sex is then XY, but the female genotype is XX, the female having two like chromosomes. In gamete formation in males, the sex chromosomes like the autosomes segregate, that is, they part from each other so that two classes of sperm are formed in approximately equal numbers: X-bearing sperm and Y-bearing sperm. Here the male is the heterogametic sex. The female, being homogametic, produces oocytes that all bear an X chromosome. It follows that upon fertilization there is an equal chance of an egg's being fertilized by an X-bearing sperm, giving rise to a female, or by a Y-bearing sperm, giving rise to a male. This explains the rough numerical equality of the sexes at conception. In some mammals the Y chromosome is absent and the genotype of the male is accordingly represented as XO.

The heterogametic sex is obviously the sex-determining sex: in mammals the male, but in birds the female is the sex-determining sex.

In mammals it is evidently upon sperm that any operation must be carried out if one sex is to be produced rather than the other. This is a matter of some importance in livestock breeding, and no effort has been spared in attempting to separate male-determining from female-determining sperm for purposes of artificial insemination. Although miracles of ingenuity have been performed, the efforts thus far have been to no avail. As to the ascertainment of sex (the first of the three meanings distinguished above), this is made possible by identification in embryonic cells of the Barr body referred to in the CHROMOSOMES entry. The ascertainment of sex is usually a bonus consequent upon the precautionary use of amniocentesis.

The mechanism of sexual differentiation is such as to yield a sex ratio of 100:100 at conception, but the sex ratio is different at birth and changes throughout life, predominantly because in utero and at all times from birth on the male is in many respects—especially in resistance to infection—more vulnerable than the female. Thus it comes about that widows greatly outnumber widowers and in general elderly women are much more numerous than elderly men. In human beings males outnumber females at birth to a degree which seems to depend upon the general state of health, age structure, and well-being of the population. Under favorable conditions (favorable in other than actuarial respects) the sex ratio may rise to as high as 106:100. It has been common experience that the sex ratio increases in favor of males during and immediately after major wars. Several explanations have been proposed, the most popular being that the especially high proportion of young brides and young mothers during and after wars, and hence the more favorable uterine environment, has the effect of sparing the more vulnerable males among the fetuses. The underlying assumption here, yet to be demonstrated conclusively, is that the primary sex ratio (the sex ratio at conception) favors males. However that may be, the effect of a relatively high sex ratio at birth is to bring about an approximate equality of the sexes during the procreative years, an equality which though highly desirable in a monogamous society is fairly often disturbed by wars, migrations, and other demographic movements.

A combination of all these factors, especially emigration of males, brought it about that in Victorian England women of marriageable age vastly outnumbered possible mates, thereby leading to a cruel excess of unmarried females: this was the great age of governesses and nannies, who were often painstakingly made to feel the supposed inferiority of their position. Sir W. S. Gilbert, whose glib doggerel was set to music by Sir Arthur Sullivan, gauged the sense and manners of theater audiences perfectly by regularly making a butt of aging, unmarried women with matrimonial aspirations. This and related aspects of his character combined to make Gilbert one of the most disagreeable figures in the history of show biz.

SIBS Together with "sibling," "sibman," and "sibness," the word "sib" (meaning a blood relation) poses something of a philological puzzle. Although it is an ancient word with recorded usages going back to the thirteenth century and although it fulfills a function not discharged by any other single word, it simply has not been received into everyday speech even indirectly, that is, through professional usage by psychologists, anthropologists, geneticists, and the like. If a child today were asked "Have you any sibs?" he or she would probably not understand the question. "Have you any brothers or sisters?" is the almost universal form. We have little doubt, though, that the word would be in common use if it had appeared even once in the King James Bible. It does appear in the Lindisfarne gospels (the English translation of the eighth century A.D.), but that is hardly enough to familiarize the English-speaking world with the word.

TWINS Unlike most mammals, human beings normally give birth to their offspring one at a time. But twinning is not uncommon, and twins are of two kinds. First there are ordinary or fraternal twins, which are in effect littermates, the products of two separate eggs separately fertilized and separately implanted into the uterus. From the way in which they are formed, fraternal twins are genetically quite distinct and, age apart, resemble each other no more closely than siblings of separate birth. Alternatively, twins may be of the kind described as identical: these derive from a single fertilized egg, as if the two daughter cells produced by the first division of the zygote had separated and developed independently. Barring genetic accidents, such twins are of necessity of the same sex. Both kinds of twinning may occur in the same pregnancy, and multiple births (triplets, quadruplets, or quintuplets) are often the consequence of both fraternal and identical twinning. Thus two siblings among triplets may be identical and the third the product of a separate fertilized egg.

The processes that give rise to the formation of identical or monozygotic twins do not invariably go to completion. If they do not, twins may be born that are not fully separated. These, through a variety of etymological misadventures, are often referred to as Siamese twins. The degree of their attachment varies greatly from twin to twin: sometimes they are conjoined abdomen to abdomen or back to back, and sometimes they share a common rectum.

Identical twinning is the only form of asexual reproduction found in vertebrates. It is exploited by the nine-banded armadillo and by the *Armadillidae* generally, for their offspring normally include a cluster of identicals derived by division from a single fertilized egg.

Ordinary fraternal twins may or may not share a placenta, that is, they may or may not be synchorial. Cattle twins invariably are. The bloods of synchorial twins often commingle freely, though of course each circulation is separate from that of the mother. When this occurs, each twin becomes a chimera: each contains its own red blood corpuscles admixed with corpuscles belonging by right to its twin.

As Francis Galton was the first to foresee, twins are profoundly useful in attempting to distinguish the relative contributions of nature and nurture to human character differences. The observations possible in theory—and sometimes also in practice—are of classic simplicity of design: the differences between identical twins who have been brought up apart from each other in different environments and exposed to different educational influences may not unreasonably be attributed to the environment, that is, in a general sense to nurture. Conversely, the differences between ordinary twins reared together and insofar as possible in a similar way are confidently attributed to differences of their hereditary makeup.

Such studies can be useful when the observations are carried out with scrupulous honesty and when valid techniques of mensuration are used.

It is especially unfortunate, therefore, that the most famous and ostensibly the most extensive of all Galtonian twin studies—those undertaken by Sir Cyril Burt—were of very questionable probity.

YOLK SAC   In vertebrates that lay shelled eggs, the yolk is contained within a double membrane known as the yolk sac. Anatomically speaking, the yolk sac is an extension of the embryonic gut. Its inner lining is continuous with that of the intestines, and its outer lining is a connective tissue richly permeated by blood vessels connecting with the embryonic heart. Yolk sacs do not have to be filled with yolk to be so described: one of the most important ways in which mammals betray their reptilian ancestry is by the possession of this sac, which plays an important part in the nourishment of the embryo and in its physiological exchanges with the mother.

ZYGOTE   A zygote is the reproductive cell formed by the union of male and female gametes, in which the regular diploid number of chromosomes is restored.

GENIUS AND INSANITY

The idea that there is a close connection between genius and insanity is a philosophical vulgarism. Its main interest lies in how it has come to be accepted into the conventional unwisdom. The best-known formulation of the idea—in which, as it seems to us, more attention is paid to meter than to sense—is John Dryden's:

> Great wits to madness sure are near allied
> And thin partitions do their bounds divide.

Dryden is said to have had the notion from Seneca. Indeed, it goes back to Aristotle and to Plato's *Divine Rapture of Creativity*. Cesare Lombroso was among the first to propose that there is a psychological correlation between genius and insanity, but Henry Maudsley stoutly denied it; the whole history of the idea is one of asseveration and counterasseveration. It is a literary rather than a psychological notion; Pascal, for example, wrote, *L'extrême esprit est voisin de l'extrême folie.*

Where all is asseveration and counterasseveration, one cannot do better than to turn to someone of strong, broad understanding;

and who qualifies better than Samuel Johnson, who in his *Life of Cowley* wrote, "The true genius is a mind of large general powers accidentally determined to some particular direction"? This was also Charles Lamb's opinion in his essay "The Sanity of True Genius." Surely he spoke with some personal authority because of the severe madness in his family; he himself periodically lost his senses.

Most of the evidence that purports to support the correlation between genius and insanity exists in anecdotes or reminiscences of men or women of enormous ability who have been neurotic or hypomanic in disposition. But if this method of inquiry is pressed to its limit, we can as easily establish a correlation between insanity and second-rate genius; men such as Friedrich Nietzsche, Robert Schumann, and Charles Lamb were talented indeed, but not to a degree that would tempt us to describe them as geniuses. Yet might they not have been so if their wits had not been to some extent deranged?

It is still difficult to understand why the notion, which seems on its face to be self-contradictory, should have caught on so readily. Perhaps it is because it seems to give a certain countenance to dullness: "I may not be very brilliant," someone of slow wits may say, "but at least I'm not out of my mind."

GERM LAYER THEORY

Formerly one of the theoretical pillars of classic descriptive embryology, the germ layer theory states that the fundamental tectonic of animal embryos is one in which the cells of the embryo are organized into epithelial sheets, cell layers, or cellular leaves or envelopes, each of which has a definite and prescribed fate in terms of the adult structures it eventually generates. There are three layers, and the first two arise as follows: when a fertilized egg has come to be subdivided into cells, the solid ball of cells usually hollows out to form a fluid-filled vesicle known as the *blastula*. In the next stage this hollow ball caves in at one pole to form two layers, the inner layer enclosing a tube, the primitive gut, which communicates with the outside through an aperture that persists as the anus. The outer of the two layers of cells is the *ectoderm,* which, being the interface between organism and environment, not unnaturally gives rise to the skin and sense organs and the entire nervous system. The inner layer, the *endoderm,* gives rise to the gut and to all organs (such as the liver) that begin their life as cellular outpushings from the gut. Between these first two germ layers various cellular movements in the embryo form a third layer, the *mesoderm,* which gives rise to the segmental muscles, the skeletal structures of the body, the musculature of the gut and body wall, and the connective tissues. The cavity within this middle layer forms the principal body cavity or coelom, from which the kidneys develop.

The great appeal of the germ layer theory rested upon its being thought that the formation of the three germ layers in development recapitulated the evolution of animals generally; thus the most primitive many-celled organisms, it was reasoned, would be hollow balls of cells like the *blastula,* and this would be succeeded by a two-layer structure such as that, basically of polyps and hydroids, forming a two-layered or "diploblastic" grade of organization. Then, as with embryos, a middle layer would be added, giving rise to a "triploblastic" grade of organization.

While the germ layer theory is no longer given much credence and the adult organs or tissues to which the various layers give rise are not as constant as they were formerly supposed to be, the element of truth in the theory is now clearly recognized. When Wilhelm Vogt developed the technique of marking areas of the embryo with harmless dyes in order to trace their movements in the course of development, it soon became clear that early embryonic development is very largely a matter of the *movement* relative one to another of cells and groups of cells. Such movements can take place only if the substance of the embryo is organized into cellular leaflets or sheets which can slide or creep over or between one another and which by outpushings and intuckings, expansions here and contractions there, lay down the fundamental structure of the adult organism. The elder Oliver Wendell Holmes (1846–1933), professor of anatomy and physiology at Harvard Medical School, likened the process to the operations of a glassblower. Thus the germ layer theory ultimately is a theory about developmental tactics rather than about developmental strategy.

GONDWANALAND

Until the last quarter of the nineteenth century the notion of the fixity of continents (that the continents always were where they now are) enjoyed the same kind of passive acceptance as the notion of the fixity of species (that species had remained where they were at the time of their divine creation). But from the 1880s on, mainly at the instigation of the Austrian geologist Eduard Suess, an entirely different possibility began to be explored: that the continents could move tangentially across the face of the earth relative to one another. This new concept was most fully embodied in what became widely known as the Wegener hypothesis, although it drew extensively on the work of Wegener's predecessors, especially Suess.

According to this hypothesis, the present land masses began as one huge continent centered on the South Pole, *Pangea;* a first fissure divided this protocontinent into a northerly part, Laurasia, and a southerly part, Gondwanaland. Laurasia comprised most of Asia and Europe as they are now, North America (including Greenland), and northwest Africa. Gondwanaland comprised the land masses that later became Antarctica, Australia, South America, most of Africa, and peninsular India. The two huge protocontinents were divided by an immense sea running in an east-west direction, Tethys (literally, a midworld sea), of which the present Mediterranean is a remnant. During the Mesozoic period, beginning more than two hundred million years ago, further fissures occurred and the continents began to drift apart to occupy their present stations.

An appraisal of the Wegener hypothesis calls for answers to two questions: did continental drift occur; if it did, how did it come about?

Some uneasiness is felt about Wegener's notion that the comparatively light rocks of the earth's crust ("SIAL," so named because silicon and aluminum are its principal elementary constituents) float upon the denser molten rocks of the earth's mantle ("SIMA" in Suess's terminology, the principal elementary constituents being silicon and magnesium).

On the other hand, the notion of drift itself is now accepted and is fully in keeping with modern concepts of plate tectonics.

The case for its having occurred does not rest solely on the jigsaw-like complementarity of structure of the Atlantic coastal outlines of Africa and South America, remarked on by Francis Bacon. The relatively new evidence from rock magnetism is strongly confirmatory and the geological correspondences across the Atlantic of Paleozoic rocks are just what one would expect if continental drift had split the Greenland coast from that of northern Europe. Drift also explains mountain formation at least as plausibly as Mikhail V. Lomonosov's surmise that mountain ranges were formed by shrinkage of the earth's crust on cooling, for the great spinal mountain ranges of the Americas, the Rockies and the Andes, might well have been thrown up when the westerly drift of North America and South America caused them to impinge upon the Pacific Oceanic plate.

There is evidence from animal distribution also, notably from "Wallace's line." This is a line of demarcation winding through the East Indies, separating Borneo and Bali from Celebes and Lombok respectively, and at the same time separating an Oriental from an essentially Australian fauna. Even more striking, because of the extreme unlikelihood that freshwater fish could possibly migrate between continents, is the fact that the existing genera of lungfish are confined to the subcontinents of Gondwanaland.

The Wegener hypothesis, the notion of the tangential mobility of the earth's crust, is one of the most audacious and imaginative innovations in the whole history of science. These are qualities which recommend it strongly to those who realize that it is through such excursions of the mind that science makes its most rapid advances; but to many lesser mortals these very qualities provide a special incentive to skepticism and to discrediting of the idea. Such men, small enough to begin with and engaged for the most part in activities that are the scientific equivalent of engraving the Lord's Prayer upon the head of a pin, feel still further diminished by any sort of imaginative brilliance. Their attitude tells us more about the nature of scientists than about the nature of science.

Perhaps, then, we should not wonder that continental drift was for many years the subject of what Lord Blackett called "violent and sometimes abusive controversy."

GREAT CHAIN OF BEING

In the entry entitled MAN'S PLACE IN NATURE we venture to suggest that the discovery of the anthropoid apes in East Africa and Southeast Asia during the voyages of discovery of the fifteenth and sixteenth centuries must have been one of the greatest cultural shocks ever experienced by mankind. Yet the writing of the period reveals none of the frightened awareness of these strange, quasi-human animals that we should expect from people who felt they had been the subjects of an awful confrontation. We look in vain for something akin perhaps to the dark, troubled thoughts of later fin-de-siècle writers like Henrik Ibsen and to a lesser extent George Bernard Shaw, upon matters to do with heredity.

What, then, insulated reflective people in about 1600 from the anxious puzzlement that should have been aroused—as it is in intelligent children today—by such a creature as that which Linnaeus later called *Homo silvestris orangoutang?* They were shielded, we believe, by the ancient conception of a *great chain of being,* or great ladder of being. In Plato's view, nothing could more clearly manifest the operations of a rational divine power than a universe in which there could be found an actual representative of any being that might be rationally imagined. Such a scheme would be imperfect if there was anywhere a gap: there must be a continuous gradation of creatures from the simplest and humblest all the way to the divine.

This was an extraordinarily vivid, persuasive, and long-lasting notion: it is perhaps the deepest root of the idea of evolution, which in a sense is the great chain "temporalized," as Arthur O. Lovejoy* puts it—spread out in time. The chain underlies the metaphysics of Leibniz and the metaphysics of Teilhard de Chardin as he propounded it in his incoherent rhapsody, *The Phenomenon of Man.*

Whatever the place of the great chain in the pedigree of thought that culminated in the evolutionary conception, it provided also a spiritual antidote to the impiety of rebutting the idea

* Everyone who writes on the concept of the great chain of being is deeply in debt to Lovejoy's *Great Chain of Being: A Study of the History of an Idea* (Cambridge, Massachusetts, 1936).

of special creation. For if every conceivable kind of organism had been specially created, would not an omnipotent Creator, answering to platonic expectations of the power of divinity, create every possible organism without gaps so that the existence of intermediate and, in the evolutionary view, transitional forms (feathered reptiles and egg-laying mammals, for example) would be seen as the expected and natural product of a great Designer?

Voltaire in his *Philosophical Dictionary* gave some thought to the great chain of being and was dissatisfied with the element in it that Lovejoy called the "principle of plenitude"—the principle that there are no gaps: all that can be imagined must be. Voltaire disagreed. Is it not obvious, he asked, that there is a gap between monkey and man? Is it not easy to imagine a featherless biped, intelligent but wordless and unlike us in appearance, which we can train and which will answer our signals and which will serve us? And can one not imagine still other intermediates between these new beings and ourselves?

Voltaire's servile mutes can indeed be imagined: H. G. Wells did so and even named them: they are the Morlocks of Wells's *Time Machine*.

GROUP SELECTION

It is usually an individual organism, in population genetics a gene, that is taken to be the subject of natural selection—that is, the element that does or does not enjoy a net reproductive advantage. We need not be so restrictive, however: any sets of replicating or replicated entities between which there are heritable differences may be the subject of natural selection. Selection of this kind is described as group selection: it is appealed to repeatedly to account for the evolutionary origin of properties that relate to the interactions of organisms, particularly those characteristics that benefit the group as a whole.

There was at one time a danger that the concept of group selection would reintroduce into evolution theory the kind of explanatory glibness that was a weakness of the older Darwinism, but modern population geneticists now stand shoulder to shoulder in repudiating the notion that "for the benefit of the species" is explanation enough for the evolution of any character trait, even if it is one that may to some degree—however small—be disadvantageous to individual members of the group or species.

A concrete example will help to make the concept plain. Consider two different populations of *staphylococci* such as might be found in neighboring hospitals. Bacteria like these are under constant and heavy selection pressures; they are not welcomed by hospital authorities, who do their best to make them feel unwanted, and they are assaulted at all times by a variety of antiseptics and antibiotics. Under such circumstances only those staphylococcal populations can survive whose members are mutable enough to throw up variants conferring some measure of resistance to one or another antiseptic or antibiotic. The minority so endowed will be rapidly expanded by selective forces until they become the prevailing type. Thus with hindsight we can say that only those bacterial populations survive whose genetic systems, especially with respect to mutability, make them versatile enough to cope with new selective pressures and hazards.

In the outcome, then, this is group selection, though what underlies it and makes it possible is of course the orthodox Darwinian selection of the individual organisms of which the groups are composed. Nevertheless, in a world in which the elements were just such groups or populations, evolution would to outward appearances be Lamarckian in style; that is to say, groups would seem to have undergone a heritable transformation by a specific adaptive response to an environmental change. It would be tiresome to press this point in the context of group selection because it is so obviously "Darwinian" at heart. Nevertheless, a distinguished physical chemist, C. N. Hinshelwood, proposed in the 1930s that if we regard a bacterial cell as a population of enzymes or of their replicating vehicles, an ostensibly Lamarckian adaptation of bacterial cells might be underlain by a straightforwardly Darwinian selection in the enzyme population of the cell, but what would the selective force be? And by what properties would an individual enzyme be advantaged in such a selective process? Hinshelwood had a good point, and if the term "group selection" had been current at the time, he would have recognized in it a certain kinship with his own ideas.

GROWTH, LAWS OF BIOLOGICAL

Two laws will be formulated here. The first is that biological growth is fundamentally of the multiplicative type; for that which results from biological growth is, typically, itself capable of growing. This property plays an important part in determining the most appropriate form in which to represent size as a function of age.

If W represents size (weight, number, or a linear dimension, as the case may be) and t represents an organism's age, then we may represent size as a function of age in the alternative ways represented by the two columns in the figure below. We may, on the one hand, plot against age the sum of all the increments of size that accumulate during life and so generate a straightforward curve of growth in which the arithmetic quantity plotted against age is $\int dW$ or, more simply, W as in (a). This curve of growth always has a point of inflection, at which the growth rate stops increasing and starts to go down. The point of inflection therefore shows up in the curve of growth rate (b) as a maximum and in the curve of acceleration (c) as its point of intersection with the time axis (acceleration = 0).

In spite of the straightforward appeal of the curve of growth and its derivatives, most students of growth have thought it more informative to plot the curve of *specific growth*. In this formulation the quantity we plot is not the simple arithmetic value W but the accumulated increments of size considered as a proportion of the size already achieved. We plot, in fact, the integral quantity $\int dW/W$, more usually written $\log_e W$. The three curves—specific growth (d), specific growth rate (e), and specific acceleration (f)—are illustrated in the right-hand column of the figure. The curves have been plotted from an equation for the Gompertz function (see the last line of the accompanying table), but the scales of the ordinates have been adjusted to make the height of each graph uniform. The table demonstrates the analytic properties of the three functions most often and most successfully used to represent the course of growth of single organisms or of populations. The last column illustrates the second of the two laws that embody the most general statements we can make about biological growth.

This second law is that the specific acceleration of growth is al-

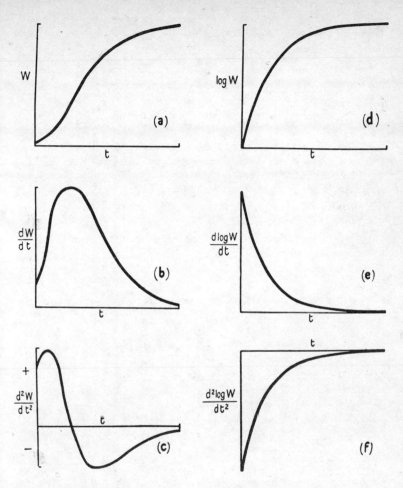

ways negative. We shall name this Minot's Law, for reasons explained in the entry on AGING. Combining the two laws into a single statement, we may say that although biological growth is fundamentally of the multiplicative type, living systems progressively lose the power to multiply themselves at the rate at which they were formed.

The sense of choosing the logarithmic plot (*d*) in the figure as opposed to the straightforward arithmetic plot (*a*) is made espe-

Function	Equation	Linear form	W $(t=0)$	W $(t \to \infty)$	Point of inflection (W, t)	Specific growth rate	Specific acceleration
"Mono-molecular"	$W = a(1 - be^{-kt})$	$\log \dfrac{a - W}{a} = \log b - kt$	$a(1 - b)$	a	—	$\dfrac{kbe^{-kt}}{1 - be^{-kt}}$	$-\dfrac{k^2be^{-kt}}{(1 - be^{-kt})^2}$
Logistic	$W = \dfrac{a}{1 + be^{-kt}}$	$\log \dfrac{a - W}{W} = \log b - kt$	$\dfrac{a}{1 + b}$	a	$\dfrac{a}{2}, \dfrac{\log b}{k}$	$\dfrac{kbe^{-kt}}{1 + be^{-kt}}$	$-\dfrac{k^2be^{-kt}}{(1 + be^{-kt})^2}$
Gompertz	$W = ae^{-be^{-kt}}$	$\log\log \dfrac{a}{W} = \log b - kt$	ae^{-b}	a	$\dfrac{a}{e}, \dfrac{\log b}{k}$	kbe^{-kt}	$-k^2be^{-kt}$

NOTE: In this table of three common growth functions and their more important analytical properties, the symbol W may be read "size"; it usually stands for weight. Logarithms are expressed to the natural base e. Many of the analytical properties may be more simply expressed as functions of W rather than of age, t; but time has been chosen as the independent variable. (Adapted from W. E. Le Gros Clark and P. B. Medawar, eds., *Essays on Growth and Form Presented to D'Arcy Wentworth Thompson* [Oxford, 1945], p. 160.)

cially obvious by the consideration that in an arithmetic scale equal divisions of the scale represent equal increments of size, whereas in a log scale equal divisions represent equal *multiples* of size.

On the face of it, the curve of growth in (a) embodies a straightforward and entirely uncomplicated concept, yet the more deeply it is probed the more complicated it turns out to be. What quantities, for example, are to be incorporated in "size," W? We must surely include finished products that are inert from the standpoint of further growing, such as hair, horn, and nail; if this is so, should not every integral curve of growth include all the hair, horn, and nail that has formed in the course of life and has been shed or worn away or otherwise lost? To avoid such embarrassing questions, some students of growth have tried to draw a distinction between a "true growth" and what they presumably think of as a "mere growth"; the multiplication of cells or of nucleic acid molecules is true, but the imbibition of water and the accretion of inert matter is mere. This usage cannot be sustained: it is a surviving remnant of the mystique of protoplasm, which distinguishes between the protoplasm of an animal, which is truly alive, and the rest of it, which is dead, so that true growth comes to be thought of as "increase of protoplasm."

Another complication arises when we ask ourselves what exactly it is that we are measuring when we plot size, W, against age, t. The curves (a) to (f) are meant to represent a typical course of growth of *one* organism, but under normal circumstances such a curve could hardly be of general interest and might be of no interest at all—unless perhaps it were the curve drawn by a doting mother of the growth of one of her young. In practice, therefore, we do not measure one organism, but more often a cohort or panel of organisms. Usually what happens is that a hundred or more organisms at birth are weighed at spaced intervals throughout life, and their *mean* weight is plotted against their age. Straightforward though it sounds, such a procedure is fraught with pitfalls, especially the danger that what are ostensibly curves of growth are in reality curves of distribution, as the following imaginary and deliberately extreme example will illustrate.

In the figure below, A is the growth curve of a hypothetical organism which about halfway through its life enjoys a sudden

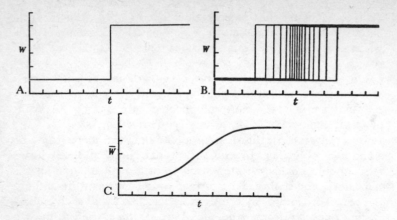

growth spurt, shown by the stepwise increment of the curve of growth. Suppose now that the time at which this growth occurs varies from one organism to another, being normally distributed about a mean value with unit variance. This state of affairs is illustrated by *B,* in which are plotted a whole assemblage of growth curves, each of the type shown by the single organism plotted in *A.* There is nothing to complain about here, except of course the extreme unlikelihood of growth's being of such a kind—something excused by the choice of the example for illustrative purposes. Suppose now we plot against age the *average* size, $\overline{W}$, of an organism by the ordinary procedure of taking averages. The "mass curve" has the curvilinear form shown in *C;* it is totally unlike the curve of growth of an individual, but according to the construction it must be a modified curve of normal distribution signifying not the course of growth but the distribution in time of the growth spurt which each individual organism has been assumed to undergo.

These difficulties and other such reservations do not detract from the significance of the two growth laws formulated above: the multiplicative law, which specifies the norm of biological growth, and the second law, which specifies the departure from the norm in real life.

HIERARCHY

A system of ordering very frequently exemplified in biology, a hierarchy is a series in which each member is included in the member that precedes it and includes all those members that follow it. A military establishment is a case in point: army, division, brigade, regiment, battalion, company, platoon, and squad constitute a hierarchy and so do the officers or hierarchs belonging to each level. Likewise, the taxons which together form a scheme of biological classification form a hierarchy: kingdom, phylum, subphylum, class, order, family, genus, species, and subspecies form a hierarchy into which any newly devised taxon such as grade, cohort, suborder, or subspecies will fit. Populations of organisms also display a compositional hierarchy, beginning at the top with a community within which there may be lesser assemblages such as geographic races and families.

Similarly, individual organisms are composed of organs built up of tissues: tissues in turn are made of cells, within which are cellular organs, composed of a variety of macromolecules which can themselves usually be resolved into simple monomers, which are of course made up of radicals and atoms—the hierarchy ending with the elementary particle. This compositional hierarchy may be brought into correspondence with the hierarchy of the biological sciences, beginning with ecology, proceeding through organismic and cellular biology, and ending with molecular biology. The hierarchy can be read either way, of course, and the order given is not to be construed as an order of merit or as an order which represents the respective practitioners' degree of self-esteem. The prevailing pecking order puts molecular biologists first, but this view is not shared by ecologists, who have a very proper sense of pride in the pursuit of a discipline which enjoys the empirical content of all the disciplines below it in the scale as well as contextually distinctive or "emergent" ideas of its own.

Because hierarchies are composed of elements which are themselves hierarchical in structure, there is a temptation to suppose that the purely formal correspondence between the various elements points to the existence of some fundamental isomorphism. A mischievous example would be to put into correspondence a political society and an organism, polymorphonuclear leucocytes then being little policemen or maybe garbage collectors, with the blood vessels the highways of communication, and the nervous system a sort of super built-in telephone exchange. The end product of this erroneous train of thought is of course the concept of the corporate state, in which there is a place for everything and everything is in its place, and in which every activity that is not compulsory is forbidden.

The temptation referred to above is one Arthur Koestler did not escape in his *Act of Creation* (London, 1964). Koestler gave a good account of the notion of hierarchy, but then went on to declare (according to P. B. Medawar, *Pluto's Republic* [Oxford, 1982], pp. 256–257)

that at each level of the hierarchy "homologous" principles operate, with the consequence that any phenomenon at one level must have its homologue or formal counterpart at each other level. In particular, we shall find "mental equivalents" of what goes on at lower levels in the hierarchy, and conversely, since we can go either up or down the ladder of correspondences, physical equivalents of what goes on in the mind. The "creative stress" of the artist or scientist corresponds to the "general alarm reaction" of the injured animal, and dreaming is the mental equivalent of the regenerative processes that make good wear and tear. Embryonic development has a certain self-assertive quality, and so have "perceptual matrices." A not yet verbalized analogy corresponds to an organ rudiment in the early embryo, and rhythm and rhyme, assonance and pun, are "vestigial echoes" of the "primitive pulsations of living matter."

No metaphors these: "they have solid roots in the earth"; but to my ears they sound silly, and I believe them to be as silly as they sound. Disregarding the rights and wrongs of building hierarchies out of non-homogeneous elements (though in fact it won't do to mix up perceptual matrices with adrenal glands, embryos, jokes and rhymes), the correspondences Koestler makes so much of are of the purely formal and abstract kind that can be expressed without any regard to their content.

HISTORICISM

An aberration of thought in the social sciences somewhat akin to geneticism in the biological sciences, historicism nonetheless deserves mention in a book that is primarily biological. Historicism stands for the belief that there exists or can be propounded a historical social science which, by uncovering the laws of social transformation (the laws, even, of human destiny), can make historical prediction as familiar and as safe a process as astronomical prediction. We must not underestimate the profound impact of Sir Isaac Newton's friend and helper, the mathematician and astronomer Edmund Halley, when he predicted in 1704 that the comet which now bears his name would reappear in 1758 (watch out again in 1986). This was a genuine marvel of science, and it was hoped that history might come to do as much if only it could master the laws of social transformation as Newton had mastered the laws of the heavenly bodies.

The justification for including an entry on historicism is reinforced by the claim made by Friedrich Engels in his oration at Karl Marx's graveside: "As Darwin discovered the law of evolution in organic nature, so Marx discovered the law of evolution in human history." (See Franz Mehring, *Karl Marx: The Story of His Life* [London, 1936].) Marx, it will be remembered, deduced the law as a consequence of the principle of class warfare which held that the condition of the working classes must inevitably deteriorate year by year. He could not have been more mistaken: so far did this prediction fall short of the standard set by Halley that it represented the very antithesis of the truth, for in industrialized countries of the West—those in which the working classes should have been most gravely at risk from the machinations of capitalism—the lot of the working classes has progressively improved, in terms of real income and also in the amenities of life and society's understanding of its indebtedness to them.

Historicism was rife in sociology until sociology found its Hume; for such is Karl Popper's position in sociological thought. In his *Poverty of Historicism,* 3rd ed. (London, 1957), and his *Open Society and Its Enemies,* rev. ed. (London, 1957), Popper undermined the explanatory pretensions of historicism as effectively as Hume had undermined the explanatory pretensions of empiricism. As a result, we hear much less nowadays of the workings of historical forces; with any luck, we shall soon not hear of them at all.

HOLISM

Holism is a doctrine that has had a powerful appeal to nature-philosophers and to amateur philosophers generally, the most famous among them being Jan Christiaan Smuts (1870–1950), former Prime Minister of the Dominion of South Africa: see his *Holism and Evolution* (London, 1926).

Holism teaches that any whole, and especially a whole organism, is not a mere assemblage of its constituent parts but enjoys an integrity or wholeness by reason of the functional interrelations and interdependences of its several parts. The philosophical opinion against which holists profess to defend us is that which is practiced by the reductive analytical-summative mechanist who believes that an organism is a mere additive sum of its constituent parts, parts among which the functional relationship is wholly explicable in physicochemical terms. It detracts somewhat from the splendor of this philosophical victory that the analytical-summative mechanist does not exist: he is a sort of lay devil invented by the feebler nature-philosophers to give themselves an opportunity to enjoy the rites of exorcism.

Holists believe (and chide mechanists for it) that mechanists study the parts of an organism in isolation and thus never understand them deeply. No complaint reveals more clearly the naivete of holists and their inexperience in practical biological matters, for the feat of studying an organ "in isolation" is in reality an impossible one. True, it is possible to remove a heart or a kidney and make it perform pumping or dialysis by transfusing it with suitable bloodlike fluids, but we are not studying kidney and heart in isolation. Their present shape and manner of working are such as they are because they have been part of a whole organism and subject to whole-organism influences from their first appearance as an embryonic rudiment to the functional finished product. There is no possibility of examining them "in isolation." Holism owes some of its popularity to the widespread belief that it is a bulwark against reductionism. However that may be, our judgment upon holism must be that if it has not advanced our understanding of biology, it has not to any important degree impeded it.

HOMOLOGY

One of the most important concepts in biology, homology was introduced and differentiated from mere "analogy" by a famous superintendent of the Natural History Department of the British Museum, Richard Owen (1804–1892). Organs A, B, and C of genetically cognate structures are homologous if the development of one can be construed as a modification or variant of the development of another, or if the developments of all three are so many variants of the development of an evolutionary precursor of them all. Thus, in spite of the differences in details of structure, the pectoral fins of fish, the forelimbs of land mammals, the wings of birds, and the wings of bats are all homologous structures; for the development of each is a variant of the development of the pentadactyl limb characteristic of vertebrates. On the other hand, the wings of birds and the wings of insects are merely analogous, being totally different in origin and development and related only by performing the same function—flight. Homology is the root concept underlying comparative anatomy.

HUNTINGTON'S CHOREA

A chorea is an involuntary-movement disorder principally affecting the limbs and the muscles of facial expression, sometimes with attendant emotional instability and mental deterioration. Huntington's chorea is a rare hereditary form accompanied by progressive mental degeneration. Its genetic background has profound implications for both eugenics and senescence, for it is the work of a dominant gene the effects of which usually are not manifest until early middle age. As a consequence, a future victim may well have had a family of ordinary size and may thus have propagated the malignant gene before he or she is struck down. A dominant and reproductively lethal gene would under normal circumstances be an easy target for natural selection, but because of its late action the force of selection against Huntington's chorea is greatly attenuated. In our entry on AGING we suggest that the deleterious genes which contribute to the syndrome of senescence become established in the population by reason of this partial exemption from the effect of natural selection.

There is no reason to think that the mutation rate of the offending gene is especially high, and the likelihood therefore is that most victims of Huntington's chorea will be the offspring of a mating between a dominant carrier and a normal spouse; approximately half their children will become victims of the disease, and there is as yet no telling which the future victims are to be. The wisest counsel that can be given to someone who because of his or her family history might be a carrier is to abstain from having children.

George Sumner Huntington (1851–1916), the New York physician who is the eponym of the disease, was said to have resolved to study it when on his rounds in Long Island he came across an afflicted mother and daughter, both cadaverously thin, twisting, mouthing, and grimacing—acting, in fact, in just such a way as to make the religious folk who abounded in the eastern states believe them possessed by the devil. This was a still graver affliction, for which the remedy in earlier times had been burning at the stake.

HYPOTHALAMUS

Between the midbrain and the forebrain is the diencephalon, the "between-brain" of which the most ventral part, anatomically speaking, is the hypothalamus—essentially the brain of the autonomic nervous system, with endocrine as well as nervous functions. The floor of the hypothalamus forms a promontory which establishes a connection between the nervous elements in the hypothalamus and the hormone-secreting element of the pituitary body. The endocrine function of the hypothalamus is to manufacture hormones, the so-called releasing hormones that activate and liberate the hormones of the anterior pituitary, especially those whose targets are other endocrine glands. In addition, the hypothalamus is the source of those hormones formerly thought to be secreted by the posterior part of the pituitary gland. These include hormones affecting the musculature of the uterus and raising blood pressure—quasi-autonomic functions. As observed in our entry on the ENDOCRINE GLAND, the endocrine functions of an organ are the functions it might well retain if its primary function were to be lost in the course of evolutionary change. There is a clear relationship between the endocrine function and the neurological function of the hypothalamus.

The hypothalamus is a center of heat regulation: the *tuber cinereum* of the hypothalamus contains temperature-sensitive cells that monitor body temperature continuously.

Localization of function in the cerebral cortex was discovered by the electrical stimulation of various points of the cortex in the course of surgical operations. In like manner, localized stimulation of the hypothalamus has disclosed the wealth of autonomic functions—and the feelings and notions that go with them—that are under hypothalamic control. These include, for example, the whole gamut of ingestive, digestive, and excretory functions. Local stimulation of the hypothalamus may bring about sniffing, licking, mastication, swallowing, and eventually defecation. Stimulation of distinct areas of the hypothalamus produces feelings of intense hunger or satiety, the former so exigent that even well-fed animals may be induced to eat voraciously, the latter perhaps opening the way to a pharmacologically rational method of dieting, in which control is not so much of eating as of the inclination to eat. It is known that destruction of the satiety center leads to morbid overeating (*hyperphagia*). Stimulation of other centers in the hypothalamus provokes fear, rage, aggression, extreme pleasure, and sleep. Certain hypothalamic injuries can lead to the distressing syndrome of which many medical students must have thought themselves victims: periodic somnolence accompanied by morbid hunger.

Study of the hypothalamus provides abundant information that is relevant to understanding the cerebral physical events that accompany acts or states of mind. Though they may help to explain *changes* in states of mind, they cannot explain the states themselves: extreme fear is no more explained by an electrical commotion in part of the hypothalamus than the sensation of redness can be explained by a train of impulses in the optic nerve that is relayed through various substations to the temporal lobes of the cerebral cortex.

HYPOTHESIS AND THEORY

In a philosophical vocabulary these words have precise meanings and correct usages. We shall attempt to expound them in order to caution biologists against such unhappy misunderstandings as that referred to in the entry EVIDENCE OF EVOLUTION— especially the idea that the word "hypothesis" has a pejorative connotation.

The declarations or propositions from which logical inferences are drawn are variously described as premises, hypotheses, axioms, or postulates. All have the same place in logical anatomy, but different words distinguish the way in which these statements came into being and the degree of confidence we feel in their empirical rightness. Needless to say, these differences show up in linguistic usages: we assert postulates, venture to propose hypotheses, assume or assert axioms, or take their truth for granted.

Hypotheses play a special role in science, for they are the portal of entry of imaginative thought. A hypothesis is an imaginative preconception of what the truth *might* be. We believe in hypotheses until we have reason to do otherwise—in Immanuel Kant's sense of the word belief ("a kind of consciously imperfect assent"). Charles Sanders Peirce (1839–1914), the foremost American logician of his era, was quite clear that without hypothesis there could be no advancement of science, and Karl Popper said the same. We cannot come to an approximation of the truth unless we begin with some preconception of what the truth might be. It is the essence of Popper's philosophy that science advances by means of a continuing duologue between *conjecture* and *refutation*, between the formulation of hypotheses and their testing by observation and experiment, as a consequence of which we either corroborate the hypothesis or are given reason to modify it. When a hypothesis is corroborated, we gain confidence in it and can, so to speak, put it in for a higher examination. In a formal logical sense hypotheses may sometimes be disproved, but there is no logical procedure by which a hypothesis can be proved true.

A theory is a hypothesis together with that which it logically entails, that is, it is a hypothesis plus the deductive inferences we draw from it. It follows, then, that a theory cannot be logically proved to be true—a point upon which Popper was perfectly clear—and so was Peirce ("the conclusions of science make no pretensions to being more than probable," he wrote). Hypotheses are, of course, imaginative in origin: we do not *deduce* hypotheses—for a hypothesis is that from which we draw deductions. Hypotheses are guesses and may of course be judicious or injudicious, foolish or wise, acceptable or not. The system of hypotheses acceptable at the moment forms what is in effect an "establishment" of received beliefs, of interpretations and theoretical constructions, that together constitute what Thomas S. Kuhn has called the prevailing "paradigm," a word that would not have caught on so readily if it had not fulfilled a need.

Nearly all scientists believe that there is a clear-cut distinction between *fact* and *theory*. The fact or the statement that embodies it is simply an affirmation of "the case," of that which *is,* purged of interpretation and theory. William Whewell (1794–1866) denied that any such sharp distinction existed and insisted that interpretation and theory are woven into our apprehension of supposedly plain facts at every level. In a finely reasoned passage ending with the words "There is a mask of theory over the whole face of nature," Whewell wrote, "Most men are unconscious of the perpetual habit of reading the language of the external world and translating as they read" (*The Philosophy of the Inductive Sciences* [London, 1837], vol. 1, sec. 4).

The construction of a theory is made possible by the existence of an underlying system of conventions or understandings that relate to the meanings of the terms used, the processes of sentence transformation, and their use in logical deduction. These constitute the *metatheory,* of which there are two main elements: logical syntax has to do with the rules of deduction; *semantics* relates to the meanings of terms and the conventions embodied in defining, that is, in the replacement of one or more terms by others. The element of metatheory which enters into science at all levels is that which has to do with *truth.* Truth is a metatheoretical concept because it is sentences or declarations that are or are not true, and any statement that a sentence is a truth or a falsity is metalinguistic—in other words, is a statement about a statement. The intuitive meaning of empirical truth is clear enough: the statement "the platypus lays eggs" is true if and only if the platypus lays eggs. The formalization of this intuitive conception in an adequately rigorous way was Alfred Tarski's greatest contribution to logic (*Logic, Semantics and Metamathematics,* trans. J. H. Woodger [Oxford, 1956], chap. 8).

The acceptance or rejection of a theory, and therefore of the hypothesis of which it represents the deductive elaboration, is very seldom a matter of merely pronouncing it to be true or false. This volume makes mention of a number of bold hypotheses: the hypothesis of evolution, the hypothesis of continental drift (see GONDWANALAND), and the hypothesis (see BLOOD AND ITS CIRCULATION) that there is an ancient, dark, and mysterious affinity between seawater and blood such that it is proper to describe blood as an evolutionary descendant of seawater.

The first two are widely accepted, but the third hypothesis is false and today is only entertained by amateurs or by naturephilosophers with a taste for obfuscation. The idea is rejected because it turns on a subsidiary hypothesis that is plainly untrue, namely that before the vertebrates left the sea, their blood or that of their ancestors had been in ionic and osmotic equilibrium with seawater. On the other hand, the theory of evolution and that of continental drift are both received theories, mainly because of their huge explanatory compass. Both do far more than explain the facts they may have been formulated to explain—facts such as the complementarity of the Atlantic coastlines of Africa and South America, the former existence of a feathered reptile such as archaeopteryx, and the present existence of a mammal that lays shelled eggs outside the body. The very least we expect of the hypothesis is that it elucidate what it was set up to explain; this correspondence does not in itself entitle us to say that it has a great explanatory compass. The enormous virtues of the theories of evolution and of continental drift are the number, variety, and unexpectedness of the phenomena they explain.

Some biological theories (we think especially of the germ layer theory and the theory of recapitulation) are definitely not true in the forms in which they are usually propounded but have remained in the repertory, so to speak, because there is an important element of truth in them; it is instructive to see how they came to take an unacceptably dogmatic form and to ask what steps should be taken to rehabilitate them. Scientific research is not a continuing clamor of affirmation and refutation, but something much more tentative and consciously exploratory in character. Outright refutation of a theory, therefore, is rather rare.

ILLNESS

Every culture, no doubt, has its theory of illness. In the Western world generally, and perhaps in all countries where a Judeo-Christian morality prevails, it is the *punishment theory of illness* that has taken hold: we are ill because we have done wrong, or sometimes because our parents have done wrong. The punishment theory is an extrapolation to disease of such homely observations as those which connect stomachache with eating injudiciously, hangover with excess drinking, and syphilis with venery.

The punishment theory is deeply erroneous and can sometimes be very cruel. The idea that prolapse of the uterus is a consequence of sexual overindulgence has already been cited as an example of Aristotle's gullibility and his habitual confusion between what *is* true and what *ought,* he felt, to be true. The punishment theory is a natural ally of the belief that sexual activity, being pleasurable, must be deleterious; not so long ago the idea that women had sexual desires and derived pleasure from their gratification was repudiated as a "vile calumny," and until recently generations of schoolboys were solemnly warned that masturbation might make them go blind and perhaps even mad.

The liberation of mankind from the dark superstitions embodied in the punishment theory of illness may be rated one of the greatest benefactions of medical science. People become ill from a variety of reasons, none of which can be construed as retributory in character: dietary deficiencies; infections by bacteria, protozoa, and viruses; malfunctionings of the endocrine glands such as those that give rise to diabetes; Addison's disease; and cretinism, the consequence of chronic thyroid deficiency. There is, incidentally, an element of charity in describing these last victims as cretins, for "cretin" comes from the French *chrétien,* meaning Christian. "The implication in this word," according to lexicographer C. T. Onions, "is that these beings are human, although dwarfed and deformed." There is precious little charity,

though, in the interpretation of congenital deformity as a punishment of the parents: it is intolerable that superimposed on the
shock and continuing misery of giving birth to an abnormal or
retarded child, the parents should be obliged to search in themselves for explanation of a misadventure that was probably due to
chromosomal accident, an unlucky conjunction of recessive genes,
or a chromosomal aberration caused perhaps by ionizing radiation.

Although in the main the theory of illness to which modern
medical science has accustomed us has been liberating and beneficent, its tendency to depreciate or obscure the influence of states
of mind on states of health must count against it. The influence of
psyche on soma has been recognized with varying degrees of conviction in resistance to some infections (notably tubercular), in
coronary heart disease, and even in cancer. It seems to us that a
doctor who pooh-poohs or dismisses altogether the notion of an
interaction between psyche and soma is not likely to be a very
good physician.

IMMUNITY

Immunity is a state of heightened resistance or accelerated reactivity toward microorganisms, transplants, or any other nonself
substances that gain access to the body. An adaptive response, immunity is differentiated from *nonsusceptibility* (in the sense that
dogs, for instance, are not susceptible to measles nor human
beings to canine distemper). Animals may be nonsusceptible to
microorganisms because they do not provide a suitable culture
medium. Thus baby rats are not susceptible to the rodent malaria
organism *Plasmodium berghei* because they do not provide enough
para-aminobenzoic acid for *P. berghei* to grow.

The present entry will serve as a memorandum of the conventional classification of immunity. It is said to be *actively acquired*
when it has been aroused by an animal's own exposure to an im-
munity-provoking stimulus. This is how most people acquire immunity to measles and other infectious diseases, and this is the
kind of immunity generated by a vaccine. Immunity is *passively acquired* when it arises secondhand: a newborn child is passively im-

munized by exposure to a maternal antibody transferred through the placenta or yolk sac or in the first milk, the *colostrum,* that a child receives from the breast. Alternatively, a vicarious immunity is conferred when blood serum from an immune subject is injected into a patient.

In professional parlance immunity is further subdivided into humoral immunity and cell-mediated immunity (CMI). Humoral immunity is that which is mediated through the action of a soluble antibody. The cells that are responsible for CMI are blood leucocytes, especially lymphocytes. An operational distinction is that humoral immunity can be passively transferred by blood serum, whereas cell-mediated immunity cannot be transferred by serum but requires transfer of cells. This is a feat not normally possible in real life; to avoid rejection of the transferred cells as foreign grafts, donor and recipient must be related to each other as closely as identical twins are. The barrier set up by the allograft reaction could of course be circumvented if CMI could regularly be transferred by means of a cell-free extract of leucocytes, a procedure which has been attempted with some success in human beings using the leucocyte extract named "transfer factor." The way it works is still unknown.

The (for immunology) characteristically infelicitous term "immunologic tolerance" is used to denote an immunologically specific state of nonreactivity toward a substance that would normally have been antigenic. The condition occurs naturally in dizygotic twins and those animals—usually cattle and sometimes human beings—in which the possession of a common placenta makes it possible for the twins mutually to transfuse each other with blood before birth. After birth such twins are found to be chimeras who can accept skin grafts and the like from each other, as sibs normally cannot. The mechanism of tolerance is not yet fully understood. Animals as a rule are tolerant of their own bodily constituents, and their failure to institute or to maintain tolerance is a cause of autoimmune disease.

The immunologic response system, we have explained, is actuated by the intrusion into the body of nonself substances. Sometimes it even comes about that the native constituents of the body behave as nonself substances. These may become antigenic in several ways: some exogenous substance may bind to the constituents of the cell and thus make them in effect foreign (which happens when industrial chemicals such as dinitrochlorobenzene or picryl chloride attach themselves to the skin and so give rise to allergic dermatitis); a second possibility is that a viral infection may alter a cell constitutionally in such a way that it acquires antigenic properties distinct from and in addition to any enjoyed by the virus itself; still another possibility is that a mutation may occur in a clone of lymphocytes and confer a new reactivity upon them—one which they would not otherwise have possessed and which, being random in origin, might by mischance confer reactivity against a native ingredient of the body.

Autoimmune disease or reactivity may arise also in a fourth and entirely distinct kind of way. As a general rule, autoimmunity is rare because, it is believed, the body acquires immunologic tolerance to its own constituents or its own secretory products. This tolerance can arise only when the immunologic response system has been exposed to the bodily constituent in the course of development, so that the body has had an opportunity to become immunologically cognizant of it. Some bodily constituents, however, may be so sequestered or may arise so late in development that this tolerance-conferring process has no opportunity to occur. Spermatozoa are an example, and so are some constituents of the lens of the eye, which is nonvascular. It is well known that an autoimmune reaction may be artificially provoked by the injection of spermatozoa and of lens protein.

Alternatively, an injury may liberate into the bloodstream what would otherwise have been a sequestered substance. Hashimoto's disease (autoimmune thyroiditis) seems to arise as a result of physical or other injury to the thyroid gland—and this process (which as explained in the discussion of positive feedback is self-acerbating for antithyroid antibodies) causes more damage to the thyroid and arouses still further autoimmunity. In fact, there may be an autoimmune element in many diseases that are accompanied by injury to bodily organs. Diseases that are suspected or known to be autoimmune in origin include myasthenia gravis, a paralysis which resembles that provoked by curare; multiple sclerosis, a progressive paralysis of stealthy origin; and certain other debilitating neurological disorders such as SSPE (subacute sclerosing panencephalomyelitis). Ulcerative colitis, sympathetic ophthalmia, and rheumatoid arthritis are also categorized as autoimmune diseases.

Because of its self-acerbating character, autoimmune disease is particularly difficult to manage. Treatment turns upon the use of immunosuppressive agents or of procedures designed to attempt to interrupt a vicious cycle (for example, washing blood corpuscles free of the plasma in which they are suspended and reinjecting them suspended in a bland plasma substitute).

The special vulnerability of the nervous system to autoimmune disease must surely be due to normal sequestration of its constituents from centers of immunologic response.

IMMUNOLOGIC SURVEILLANCE

A notion that embodies a proposed teleology of the allograft reaction, immunologic surveillance attempts a rationale of the immunologic process that normally prohibits transplantation of organs and tissues from one individual to another. The argument is that the rejection of allografts is a tiresome by-product of the existence of a body-wide monitoring system, of which the primary purpose is to identify and cause the rejection of malignant variants of normal tissue cells—of incipient malignant growths, in fact.

This attractive, plausible, and widely accepted hypothesis was first propounded by Lewis Thomas and developed by Macfarlane Burnet; the will to believe it to be true has always been stronger than the evidence in its favor. Its acceptance is of course closely related to the immunologic theory of natural defense against cancer. Pathologists give widespread credence to the thesis that incipient cancers are far more numerous than those that eventually graduate to become a clinical threat: the growths that do not progress, it is argued, must have been suppressed by some normal mechanism of surveillance against cancer.

The hypothesis is certainly not true in the rather naive form in which many of us at first accepted it—a form based upon the supposition that incipient tumors are eliminated by an immunologic process almost exactly akin to that which brings about the rejection of allografts. This may only mean that a cell other than the peripheral blood lymphocyte is the principal agent of surveillance: an alternative possibility is the complicity of the so-called natural killer cell. The hypothesis must stand or fall with the immunologic interpretation of resistance to tumors: if that hypothesis is false, then the idea of immunologic surveillance must surely be mistaken.

Right or wrong, the idea has fulfilled the principal function of a scientific theory: it has led to observations and experiments that would not otherwise have been performed, in this case substantially enlarging our overall understanding of cancer.

INDIVIDUALITY

A pseudoproblem invented by nature-philosophers to give biology an air of profundity, the "problem" of individuality arose out of the following circumstances. Several groups, of which representative members are solitary and sessile and which accordingly have considerable powers of regeneration and of asexual reproduction, exist in colonial as well as solitary forms. Examples are corals and salpas and pelagic organisms such as *Physalia*, the Portuguese man-of-war, and its relatives. These colonies are, so to speak, biological polymers, of which the structural unit or monomer corresponds in structure to the solitary sessile form characteristic of the group. In corals this is an individual hydroid or polyp, and in salpas it is a sessile individual tunicate. The so-called individuality problem is whether these colonies (which are highly integrated structures with a good deal of division of labor among the individual parts) should be regarded as colonies or as individual organisms.

When we authors were students, the individuality problem was one of the topics upon which we were required to write critical and analytical essays. We were not sophisticated or philosophically literate enough at the time to denounce the individuality problem as a fraud—a nonproblem of very little interest, upon the answer to which nothing of any importance turned. For an organism such as *Pyrosoma* is both an individual *and* a colony compounded of individuals: there is no ruling that an individual organism, properly so called, must consist only of individual cells. In many respects a termite colony is an individual organism—a sort of social polymer—of which the constituent parts correspond morphologically to the individual solitary forms.

Nature-philosophers had an unhappy propensity to invent problems. Another pseudoproblem was the question of whether connective-tissue fibers are alive, as cells are alive, or are to be dismissed as dead. The manufacture of all such problems is entirely supererogatory. Biology is already difficult enough. A genuine problem having to do with individuality is discussed in the entry on TRANSPLANTATION.

INFERTILITY AND ITS REMEDIES

Infertility in marriage is much more widespread than is commonly thought: reliable estimates put its frequency in the United Kingdom today at about 10 percent. For some causes of infertility manipulative remedies have been devised that may involve fertilization of the female egg cell in vitro (literally, "in glass"—in effect, in a suitable vessel outside the body) before implantation into the mother's womb. Female infertility sometimes results from an obstruction or other abnormality that prevents an egg cell from passing down the oviduct into the uterus, in the wall of which, after fertilization, it normally becomes implanted. A remedy that has been recently introduced is the in vitro fertilization of the oocyte with the husband's sperm. Reimplantation of the fertilized egg into the uterus may be followed by a normal pregnancy and delivery.

Babies raised by this procedure have been referred to as test-tube babies—a highly injudicious misnomer in which a taste for alliteration has got the better of common sense. These are not babies in any ordinary use of the term: they are *embryos*—at this stage lacking brain, eyes, ears, or nervous system. To describe them as human beings because they derive from and turn into human beings is on a par with describing an acorn as an oak. Embryos, babies, and human beings are different in fact and accordingly in name: to use for one the name appropriate to one of the others is a widespread cause of mischief and misunderstanding.

In order to improve the chances of successful fertilization in vitro the mother may be induced by hormone treatment to "superovulate"—that is to say, to produce several egg cells at the same time. Inasmuch as only one of these will be fertilized and used for uterine implantation, the remainder, if not discarded, are sometimes used for the study of human development—for example, stimulating a fertilized egg cell into forming a clone. The discarding of unused egg cells has been likened to murder by opponents of the whole procedure, on the grounds that the object being experimented on is a living human being. The general public, however, supported by a sturdily Johnsonian common sense, regards the process as a humane and gratifying remedy, of which the effect is to bring into the world a child enjoying one of the most impressive advantages a newborn babe can have: that of being very dearly wanted.

When the cause of infertility is some deficiency or inadequacy on the part of the male, different expedients may be used, especially those involving the use of AI (artificial insemination), a procedure which the media do their best to represent as appetizingly salacious in character.

When a male partner can produce ostensibly normal sperm but for one reason or another, whether psychological or physical, cannot introduce them into the vagina, artificial insemination by the husband's sperm (AIH) may well be resorted to. If things go well, the child so conceived is in every sense its parents' child. This procedure is not normally objected to on moralistic grounds, presumably because the entire operation is too clinical in character to arouse prurient interest.

On the other hand, there has been strong opposition to AID, artificial insemination which uses the sperm of a voluntary donor. The primary objection has been that the parental relationship under these circumstances is not genuine and that the child is no more than a step-child. It has been objected, too, that if AID were to become widely prevalent, we should populate the world with half-brothers and half-sisters and thereby risk inadvertent inbreeding, with the ensuing genetic damage.

The likelihood of such a misadventure is negligibly small, and we are not impressed by the idea that artificial insemination is likely to supplant the natural procreative processes and parental feelings that have sufficed hitherto. As to AID's being "unnatural," we regard the children born in this way as adopted children, with the advantage over ordinary adopted children of having grown within and having been delivered by the adoptive mother, who may accordingly be expected to feel particularly close to them.

All the procedures we have discussed are special remedies for special ills: ordinary folk can be pleased and reassured by the thought that such antidotes are now available to them—unless indeed they are withheld out of religious bigotry or because of the strong but usually unfocused resentment that not very bright members of the laity feel at the supposed usurpation and diminishment of human prerogatives by modern science. We find this a ridiculous view, for all the procedures we have described encourage a restoration rather than a withdrawal of human prerogatives.

INSTINCT

Instinct is a concept more easily described, explained, and exemplified than made the subject of a formal definition. For the benefit of those insecure people who feel that no discourse is possible unless words are defined—a counsel of despair nullified by the fact that biologists have discussed instincts for many years without attempting a formal definition—the term "instinct" is almost invariably used to describe not a single action, but a functionally connected train of performances having what ordinary people describe as a purpose (for example, a newborn marsupial's spectacularly precarious climb from womb to pouch).

Perhaps the first property that comes to mind when the term "instinctive" is used is the property of being unlearned—of being ostensibly the acting out of an innate or an inherited scenario. It should be added, however, that some performances described as wholly instinctive in reality depend, as bird song does, on environmental cues. Automatism is another characteristic traditionally associated with instinctive behavior, especially its performance even when it is functionally inappropriate and contra-adaptive, and it is often combined with a certain inflexibility that prohibits modifications to suit special circumstances.

As a figure of speech, performances are sometimes described as instinctive when all that is intended is reference to the functional aptitude of a performance combined with its prompt and effective execution. Thus accomplished drivers sometimes describe their driving as pretty much "instinctive," a usage that illustrates very well the truth of Alfred North Whitehead's wise judgment in his *Introduction to Mathematics* (London, 1911, p. 61): "It is a profoundly erroneous truism, repeated by all copy books, and by eminent people when they are making speeches, that we should cultivate the habit of thinking what we are doing. The precise opposite is the case. Civilization advances by extending the number of important operations which we can perform without thinking about them." We regard this as a profound observation which most people will be able to bear out from their own experience. Learning a new repertoire of neuromuscular performances such as driving a car or skiing is precisely learning *not* to be obliged to think about them. It is learning to react promptly and appropriately to each stimulus as it presents itself. Our behavior must be quasi-instinctual, for these are performances in which ratiocination has no part, though judgment—it is to be hoped—will have been exercised by insuring against the breakage of one's limbs that so often accompanies skiing and other people's misjudgments when driving a car or, as English people absurdly say, "motoring."

The neuromuscular scenarios of instinctual performances in the social behavior of insects and mating behavior of animals are extraordinarily complex and at present we have no inkling of how they are programmed. Still more extraordinary is the situation in which a behavioral program is merely latent, although some appropriate stimulus may call it forth. Such is the case, for example, among chickens, for cocks may be caused to display the entire repertoire of female parental care if they receive injections of female sex hormones. The more one thinks about this circumstance, the less easily explicable it seems to become. Biology has no greater triumph to look forward to than a solution of the problem of how a program of instinctual behavior is genetically stored and epigenetically retrieved.

What about plants? To describe any activity of a plant (such as the astonishing performance of an insectivorous plant like

Venus's-flytrap) as "instinctual" violates natural usage, yet the word would have been used if plants had nerves and muscles. In spite of the fact that a plant's "psychism," in the words of Teilhard de Chardin, is so "diffuse," there is no reason why biologists should regard the possession of a neuromuscular system or a psyche as a necessary condition for an organism's behavior to be described as instinctual. All that need be said is that plants have effectors of their own and sometimes perform in ways which it is not fanciful—and might even be instructive and liberating—to describe as instinctual. Plant behavior is a subject that needs to be more fully explored.

INTERFERON

It is well known that the infection of a cell by one virus makes it relatively refractory to infection by a second virus. The chemical agent shown by Alick Isaacs and Jean Lindenmann in 1957 to be responsible for this phenomenon of viral interference is known as "interferon," or, more properly, "the interferons." Although interferon is not a single homogeneous substance, interferons have a variety of natural functions and clinical uses in local or systemic virus infections, and for reasons still unknown are useful in the treatment of certain forms of cancer—even if the cancer is not or is not known to be of viral origin.

The full capabilities of the interferons are not yet known, for they are species specific and work best in members of the species in which they themselves were formed. For this reason interferons are in chronically short supply. Only when adequate amounts become available—quantities on the order of kilograms—can their full potentialities be investigated on an adequate scale.

It is helpful, though, that interferons can be produced by other than viruses—by, for example, a variety of so-called interferon inducers, among them a number of the double-stranded synthetic polynucleotides. The production of interferon has been attempted in the following ways. First, a human leucocyte concentrate or cultured human cells, sometimes cancerous cells, can be treated with viruses or interferon inducers. The yields, however, are adequate only for small-scale laboratory experimentation. A second procedure is to introduce interferon genes into microorganisms by the techniques of bioengineering, and then culture the microorganisms on a massive scale. The third method, which shines before us like the Holy Grail (and like the Holy Grail seems sometimes to recede as we approach it), is organic synthesis—a formidable task indeed: although natural interferon molecules have carbohydrate attachments, the working moiety of the molecule is thought to be a protein with a molecular weight of about 4,000.

But these production problems are soluble and will be solved; they are a necessary condition for future progress. When sufficient quantities are available to expose interferon to all the tests laboratory scientists and clinicians have in mind for it, we are in for some very pleasant surprises. In our judgment interferon may well come to rank as one of the truly impressive biomedical discoveries of the twentieth century.

INVERTEBRATA

A word that has the ring of a taxon, "invertebrata" is in reality merely a description of a miscellany of animals united only by the property of not possessing a backbone. Clearly the *lack* of a characteristic will not in itself carry taxonomic weight, for if it did protozoa, also lacking backbones, would be rated invertebrates. And so would plants. Convention therefore restricts the use of the term to metazoa, to many-celled animals.

The groups that constitute the invertebrates are distinguishable into two major lines of evolutionary descent, the distinction being based partly on embryonic development and partly on the fundamentals of adult structure. Thus the annelids, the arthropods, and for developmental reasons the mollusca form one such line; the annelids and arthropods possess a ventrally placed nervous system and a dorsal heart. To this lineage we may add other groups almost certainly derived from it by neoteny, such as the roundworms (nematodes) and the so-called wheel animalcules, the rotifers.

Vertebrate animals characteristically have a dorsally placed nervous system and a ventral heart. Several invertebrate groups are united by lying on or close to the major line of descent that leads toward the chordates. These poor relations are the *Echino-derms* (sea urchins, starfish, and sea cucumbers) and the group of which the principal member is amphioxus; in addition, there are a few minor groups whose claims for admission to the chordate line of descent are in some ways less convincing—the groups of which the representative members are arrowworms (*Chaetognatha*), the *Phoronida,* and moss animals (*Ectoprocta*).

IRRITABILITY

A term that almost invariably figures in lists of differences between living and inanimate matter, "irritability" means no more than reactivity or responsiveness to changes in the environment. Certainly the complete lack of any such responsiveness is often construed as evidence of lifelessness.

"Irritability" has, however, a more particular meaning: Johannes Müller, the great German physiologist (1801–1858), formulated a Law of Specific Irritability, sometimes called his Law of Specific Nervous Energy, which we describe in the entry on SENSE ORGANS.

KING CRAB

More than one crablike creature is referred to as a king crab, the most famous being the ancient (at least two hundred million years old) "horseshoe crab," *Limulus*.

Victims of the mysterious neurosis known as *arachnophobia* will doubtless shun the king crab when they learn that, far from being a regular crustacean, the king crab belongs to Arachnida, a group closely related to the spiders. This classification was first divulged to the world by that professor of zoology at University College, London, who is said to have been the model for Sir Arthur Conan Doyle's Professor Challenger in *The Lost World* (London, 1912).

"Limulus, an Arachnid" was the peremptory title of a famous paper written by Sir Edwin Ray Lankester. John Collier once painted Lankester lecturing on Limulus and looking at a dried specimen as if he feared it might answer back.

Another professor of zoology, not a member of a school noted for deep evolutionary learning, attempted to derive the *Vertebrata* from forms closely akin to horseshoe crabs, a feat of legerdemain rather similar to that which musicologists sometimes use when demonstrating that one musical theme can be derived from another: all that need be done, they sometimes explain, is to play the tune backward or upside down and alter the tempo from waltz time to 5:4 and presto, the point is made.

LAMARCKISM

"Lamarckism" was not Lamarck's term, any more than "Darwinism" (coined by Alfred Russel Wallace) was Darwin's. Although not many people today bother to acquaint themselves with the evolutionary opinions of Jean Baptiste Pierre Antoine de Monet, le Chevalier de Lamarck (1744–1829), no one is in much doubt about what Lamarckism connotes. It epitomizes the belief that changes acquired during an individual's lifetime as the result of active, quasi-purposive, functional adaptations can somehow be imprinted upon the genome, thereby becoming part of the heritage of succeeding generations. Very few professional biologists believe that anything of the kind occurs—or can occur—but the notion persists for a variety of nonscientific reasons.

1. It has been the subject of persuasive propaganda by literary figures from Samuel Butler to George Bernard Shaw to Arthur Koestler. Butler believed that instincts are genetically encoded habits. Shaw made some characteristically tiresome comments on the subject, doubtless inspired by the thought that he was having a go at the "establishment." Koestler was not a Lamarckian, but in his famous book on the Kammerer fraud (*The Midwife Toad*) he created an atmosphere favorable to the acceptance of Lamarckism by representing one of its principal proponents as a victim of establishment orthodoxy and also by his criticisms of neo-Darwinism (of which he did not give a very expert or just account).

2. Marxists scorn the Darwinian notion that there are inborn differences of human capability and choose instead to believe that men are born equal and become what their upbringing and environment make of them, a notion Napoleon Bonaparte is known to have thought abhorrent. It is not surprising that Lamarckian thought had an influence on French revolutionary thinkers, nor that it was official doctrine in Russia at the time (roughly 1937–1964) when agricultural science was dominated by the

opinions of Trofim Denisovich Lysenko and Mendelian geneticists were in disgrace for believing in the genetic inequality of man.

3. It seems only natural justice that an animal's endeavors and the outcome of its own exertions should become part of its genetic heritage.

4. Many naturalists have lost faith in what sometimes has seemed to be the all-encompassing explanatory glibness of neo-Darwinism.

5. The form of social evolution or inheritance that is mediated through culture (discussed in the entry on EXOGENETIC HEREDITY) is clearly Lamarckian in style.

6. Until very recent years it was widely believed that evolutionary changes in bacteria and other microorganisms, such as those which accompany what is revealingly described as the "training" of microorganisms to utilize new foodstuffs or resist the action of new antibiotics, are Lamarckian in style. This misapprehension, as it turned out to be, gave Lamarckists a special incentive to try to find a similar process at work in higher organisms.

For all these reasons, claims that Lamarckian inheritance has been demonstrated recur regularly, cause at first a flurry of excitement, and then drop from sight. We shall cite a few examples and ask the reader in appraising them to remember that no Lamarckian interpretation can be accepted as valid unless it expressly excludes the possibility of natural selection.

Industrial melanism in moths. It has been said that "the spread in industrial districts of melanic forms [that is, of dark or dusky variants] of Lepidoptera is ... one of the most considerable evolutionary changes that has ever actually been witnessed." According to Heslop Harrison, melanism is an induced and inheritable adaptive change brought about by the contamination of food plants in industrial areas by metallic fumes which were said to be capable of inducing the formation of melanic mutants. There is, however, some doubt that salts of manganese or lead can in fact induce such mutant formation. As there is no doubt whatsoever that melanic mutants occur regularly in many species of Lepidoptera in nonindustrial areas, an explanation in terms of natural selection is now thought to be much more likely: melanic mutants enjoy a distinct selective advantage amid the discolored foliage of an industrial countryside.

Inheritance of learned behavior patterns in rats. Learning may be regarded as a purposive adaptive response and may thus qualify to be considered as an adaptation eligible for Lamarckian inheritance if it occurs. In papers published over a period of years in the *British Journal of Psychology,* William McDougall described experiments on the inheritance of the ability of rats to learn to use one of two alternative exits from a water trap: one exit was brightly lit and arranged to give any rat that used it a sharp electric shock, the other exit was dimly lit but not electrified. Learning status was quantified by assessing the number of tests that had to be applied before an individual made twelve correct choices of exit successively. The rats that had been bred, as far as possible unselectively, from the tank-trained population improved in performance from a score of 120 to a score of only 36 in the thirty-fourth generation. It looked as if a prima facie case for Lamarckian inheritance had been made. It was not so simple, however: rats belonging to two different control groups also improved in performance, the controls being either untrained rats or "negatively selected" rats, bred from the dullards in each generation. Indeed, the dullards improved more dramatically than the tank-trained rats.

The errors to which experiments of this kind are vulnerable would have been far from obvious to experimental psychologists. Although they look alike (see ALBINISM), Wistar rats are flagrantly heterogeneous—that is, genetically nonuniform. Differences of intelligence in rats demonstrated by maze performances are highly susceptible to selection, and it may well be that selection was inadvertently practiced (even, for example, by so innocent-sounding a device as choosing the parents of the next-generation rats from the largest, healthiest-looking, or most active litters or from the first to breed). The improvement in performance of the negatively selected group of rats would suggest that inadvertent selection was in progress—perhaps selection for heterozygosity, and thus hybrid vigor, in a partially homozygous stock.

However that may be, careful and extensive repetitions of McDougall's research failed altogether to confirm it. His work therefore becomes an exhibit in the capacious and ill-lit museum of unreproducible phenomena.

Inheritance of eye defects induced by the action of antilens antisera. Between 1918 and 1934 M. F. Guyer and F. A. Smith described experiments in which they professed to show that the injection into pregnant rabbits of an antiserum made by injecting lenses of rabbits into ducks caused defects in eye development that were transmitted through both male and female lines in the manner of the Mendelian recessive. These experiments are plausible in some ways and extremely implausible in others. As an embryonic target, the lens was a good choice because antibodies directed against any other organ would almost certainly have been removed from the circulation by the pregnant mother's tissues, but antilens antibodies would not have been. (The lens of the adult is not a vascular structure, so antibody would not reach it.) On the other hand, it is not very likely that a passively transferred antibody made in another species (to wit, ducks) would be able to penetrate the yolk sac of the fetus, the principal route by which antibodies enter the rabbit's embryo; moreover, congenital defects of the eye are not uncommon in domestic animals, and if by chance one turned up in a rabbit that had received injections of antiserum, it would very likely have been inherited according to Mendelian rules.

For these various reasons it is not surprising that independent repetitions of the work of Guyer and Smith did not confirm them.

A common objection to Lamarckism is that there is no conceivable method by which an adaptation acquired in an individual's own lifetime could become imprinted upon the genome. Although it is now known to be possible for RNA to be mapped under certain circumstances on DNA (by means of the enzymes known as reverse transcriptases), there is not a shadow of evidence that the mapping of RNA on proteins can be put into reverse. It is of course quite irrelevant that mutation can be induced by an external agency, notably by ionizing radiations such as X rays—irrelevant because there is no functional or adaptive relationship between the character of a mutant and the nature of the mutagen that caused it: mutations do not arise in response to an organism's needs nor do they, except by accident, gratify them.

For the present we have no alternative but to conclude that Lamarckism is not an intellectually viable doctrine and that, whatever its shortcomings, we shall probably have to live for many years with Darwinism.

Lamarckian inheritance seems to those who believe in it to be so obviously, so necessarily, and so self-evidently true that they can develop a strong sense of grievance against nature when it does not cooperate with the experimenter who sets up a system intended to demonstrate the inheritance of acquired character. This sense of grievance has at least once taken the form of helping nature to produce the right results, and quite often the form of scathing personal attack on biologists who refuse to accept Lamarckian theory until satisfactory evidence is brought forth ("satisfactory evidence" being evidence that is not open to alternative and orthodox interpretations). Refer to the LYSENKOISM entry for evidence of the tragic consequences of the fanaticism that so often accompanies Lamarckian beliefs.

LANGUAGE

Only human beings communicate with one another by a "true" language—that is to say, by vocalizations that may have abstract meanings and are subject to inflections and variations of syntax such as may distinguish between tenses (past, present, or future) and between moods (indicative or subjunctive).

While no other animal competes for the title of being a "language animal," at the same time there can be very few animals that have no means whatsoever of communicating with fellow members of their species—whether by rude vocalizations, gestures and grimaces, dances or other bodily movements. Foraging bees reputedly convey the direction and distance of provender to their hive mates by stylized flight patterns and dances. Although the specificity of the information so conveyed has been called in question, critical experiments have been reassuringly confirmatory. Philologists reject with undue asperity the suggestion that these elementary communications constitute even the most primitive rudiments of language, but we feel that they are a little too nice in their judgment: it is as if they were to say that a childish utterance like "Johnny go walkies" is too far removed from true language even to rate as a linguistic utterance.

It is an implication of the evolutionary canon that even the most complicated behavior must have a beginning, and it would not be surprising if in lower anthropoids this should take the form of grunts and grimaces, the latter made possible by the fact that the higher primates have faces—a distinction of almost taxonomic stature. Language cannot have begun in the form it was said to have taken in the first recorded utterance of Thomas Babington Macaulay (the infant Lord Macaulay): once when he was taken out, his hostess accidentally spilled hot tea on him. The little lad first bawled his head off, but when he had calmed he said in answer to his hostess' concern: "Thank you madam, the agony is sensibly abated." It is not likely that this account of the matter is literally true; no more is it likely that our human language sprang into being in some form that a philologist would admit to being truly linguistic.

LEMMINGS

Rat-sized, stumpy-tailed rodents of the Arctic or north temperate zones, lemmings are noteworthy for their remarkable fluctuations of population size and for having given rise to a myth remotely akin to that of the Gadarene swine, namely that lemmings control their numbers by a public-spirited suicide in which they cast themselves over the cliffs into the sea. One would have to go back to the travelers' tales of the seventeenth century, or even to Aristotle, to light upon another misconception as comically erroneous. The impossibility of a genetically programmed suicide is made clear by the entry ALTRUISM.

LIFE TABLE

Imagine a large cohort—say 100,000—of human beings to be labeled or otherwise identified at birth. Imagine also that a record is kept of the ages at which each member of the population dies, until the table is completed by the death of the last one. Such a table might well be called a "death table," but by a happy euphemism that makes it easy to understand why demography is often classified among the humanities rather than the sciences, it is in fact called a "life table"; and a life table is the actuary's most important tool.

By convention, the number of the cohort with which the life table starts is designated L_0 and the number surviving to age x is L_x: obviously L_x declines as x increases. From the life table a number of demographic characteristics of a population can be read off directly: thus the likelihood at birth of living to age x is L_x/L_0, and in general the likelihood at age a of reaching some greater age L_b is given by L_b/L_a. Further, the death rate is the rate of decline of the number of survivors, that is, $-dL_x/dx$, the minus sign giving the resultant figure a positive value. More instructive than the death rate, however, is the age-specific death rate, or force of mortality—the death rate at any age of life expressed as a function of the number of people who have lived that long. This is sometimes called μ_x and is given by

$$- \mu_x = \frac{1}{L_x} \frac{dL_x}{dx} \text{ or, identically, } d \ \frac{\log_e L_x}{dx}.$$

Life tables may be constructed not only for living organisms but for any material object which suffers something equivalent to mortality—for automobiles, restaurant crockery, and the like. The actuarial characteristics of such a life table can give a lot of information about causes of wastage and the desirable pattern for replacement of losses; demographic statistics on automobiles would also show whether it was always judicious to buy a new car or whether in some circumstances it would be better to buy a car already a year or two old.

LUNGFISH

Lungfish are river-dwelling or lake-dwelling fish which, as their zoologic designations *Dipnoi* and its alternative *Dipneusti* imply, have two respiratory resources. In addition to "breathing" dissolved oxygen by means of gills as other fish do, they can breathe atmospheric oxygen by means of lungs that arise as modifications of their swim bladders, which acquire an alveolar structure essentially similar to that of lungs. A variety of interesting vascular modifications goes with this unusual habit: there is some functional separation of pulmonary and systemic circulation, and there is a coronary arterial circulation as well. The drying of riverbeds and lake beds during the summer months leads the lungfish to a state of dormancy or aestivation at which time pulmonary breathing takes over.

The ancestors of the animals intermediate between fish and amphibians must have been in many ways like modern lungfish, thereby illustrating the bridge between aquatic and terrestrial environments.

The geographic distribution of today's lungfish (*Lepidosiren* species in South America, *Protopterus* species in Africa, *Neoceratodus* in Australia) is wholly consistent with the notion that the modern continents were derived from a single precursory landmass, Gondwanaland, by fission followed by continental drift.

LYMPHATICS AND LYMPH NODES

The word "lymph," rather loosely used in physiology, has a technical meaning that should be adhered to: the word denotes the fluid content of the vessels, the *lymphatics,* that drain the tissues. Lymph arises as a filtrate of blood and is conducted via the lymphatics back to the venous system. Lymph plasma is very like blood plasma in composition, and like blood plasma it will form a clot. Under normal circumstances red blood corpuscles are not present in lymph, but cells of lymph are various kinds of lymphocytes (or cells that give rise to them such as lymphoblasts).

The lymphatic system begins as capillaries that permeate most tissues—though not the brain—and unite into lymph veins, which are thin walled, easily collapsible, and (contents apart) distinguished from other veins by possessing a remarkably large number of valves. These valves prevent the lymph from flowing in any direction except *away* from the tissues in lymph veins of increasing diameter that eventually open into the venous system.

The hydrostatic pressure of blood is not nearly high enough to cause lymph to flow: its one-way circulation is brought about by the squeezing of the lymphatics through muscular contraction or bodily movements generally. Massage is an effective way of getting lymph into motion, and experienced masseurs act as if they know instinctively the layout of the lymphatic vessels and the way to propel fluid along them. Some vertebrates—frogs among them—have contractile vessels ("lymph hearts") somewhere in the lymph circulation, but generally speaking lymph is moved passively as a by-product of bodily movements.

Usually one or more lymph nodes lie in the pathway of the lymphatic vessels and filter the lymph on its way to the venous system. A lymph node is very often the first site of engagement of an antigen and the cells that respond to it immunologically. When the antigenic matter consists of bacteria, there is frequently a decided inflammation of the nodes (usually referred to by laymen as the lymph glands): these are what a physician palpates when examining a febrile patient.

Although they have an important part in mounting immunologic responses, the functions fulfilled by the lymph vessels are not wholly beneficent, for it is through the lymphatics that tumor cells are so often disseminated through the body. (Here, too, lymph "glands" may swell because of the proliferation of tumor cells within them.)

LYMPHOCYTES

Lymphocytes, as remarked in the entry on BLOOD AND ITS CIR-CULATION, constitute a genus of white blood corpuscle—the ones that play the most important part in the transaction of immunologic reactions. Although they have been known as microscopic objects for more than a century, the understanding of their behavior and properties is the work of comparatively recent years.

The authors of this book carried out their graduate studies in Oxford University's School of Pathology, at the time presided over by H. W. Florey, whose principal work in those days before the advent of penicillin was to discover the life history and functions of lymphocytes. We were told of and invited to ponder upon a statement by one of America's foremost pathologists, Arnold Rich, who had declared, "Our complete ignorance of the function of the lymphocyte is one of the most humiliating and disgraceful gaps in all medical knowledge." We were made acquainted, too, with what came to be called "the great lymphocyte mystery." It had long been known that the principal lymphatics united to form lymph veins of increasing size and that these finally poured their contents into the great veins of the neck through left and right thoracic ducts. By this means something of the order of 10^9 lymphocytes entered the bloodstream every day. The great lymphocyte mystery was, what did they do and what happened to them? Experiments making use of radioactive labeling and studies on the longevity of chromosomally aberrant lymphocytes in the blood of patients who had received high doses of x-irradiation have since made it clear that far from being, as they had been thought to be, short-lived cells which disappeared from the bloodstream by simply dying, most lymphocytes had lifetimes on the order of months or even of years. This merely deepened the mystery.

The solution, appropriately enough propounded by a gifted experimentalist working in Florey's laboratory, was that the lymphocytes in the blood at any one time left the blood via the lymph nodes, reentered the lymphatics, and in due course were carried into the blood again. In short lymphocytes *circulate*, resembling in this respect members of the chorus in a provincial production of the opera *Aida,* who also reappear after passing temporarily from the scene. The experimental pathologist who discovered and described this crucially important phenomenon— Sir James Gowans, currently head of the Medical Research Council of Great Britain and Northern Ireland—is the same person who, with the collusion of many others, discovered that the function of lymphocytes is immunologic.

Lymphocytes were described above as constituting a genus of cells, for the very good reason that they are of many different species. Some are antigen sensors, specially adapted to recognize nonself substances that gain access to the body parenterally. Others are effector cells of alternative types—either antibody-forming cells or the agents of an immunity of the kind classified as cell mediated; the latter type may make contact with target cells and destroy them. Still others play a supporting role in the presentation of antigen to responding cells or in promotion of the proliferative response to which such an exposure normally leads. It is very important that no significant fraction of lymphocytes are merely bystanders, like the crowd that collects at the scene of an accident. The presence of lymphocytes in the rejection of grafts, in tumors, or in autoimmune diseases such as rheumatoid arthritis is invariably a signal of events related to immunologic response.

LYSENKOISM

The scheme of beliefs known as Lysenkoism was named after Trofim Denisovich Lysenko (1898–1976), the evil genius of Russian genetics and agrobiology who singlehandedly arrested the teaching and practice in Russia of "Mendelist-Morganist genetics" (that is, of genetics) and brought about the disgrace of its principal practitioners.*

The motives that led Lysenko and his colleague, the lawyer I. I. Prezent, to denounce Mendelism and Morganism as metaphysical, idealist, reactionary, and an agency of class warfare were probably twofold: political expediency and a jealous detestation of the leading Russian geneticist and agrobiologist, Nikolai Ivanovich Vavilov (1887–1943). Political expediency came into play because the genetic inequality of man and the idea of selection based upon inborn differences were deeply repugnant to official Marxist theory—an error of judgment for which Marx himself cannot be blamed. Jealousy of Vavilov's distinction and the esteem in which he was held throughout the world was enough to arouse the animosity of such as Lysenko. Moreover, it was Nikolas Maksimov, a worker at Vavilov's Institute of Plant Breeding, who detected and first described the phenomenon of vernalization (a discovery Lysenko had arrogated to himself). This procedure involves the brief exposure of seed to cold and thereby makes it possible for certain winter wheats to be sown in spring for summer harvesting. Vernalization was cited by Lysenko as proof positive of the power of environmental influences to produce specific directional changes in the genetic makeup of organisms—the naive and indeed erroneous belief characteristic of Lamarckism.

The deep-seated lack of clarity of the theoretical views of Ly-

* The standard reference work is Zhores A. Medvedev, *The Rise and Fall of T. D. Lysenko,* trans. I. Michael Lerner (New York, 1969).

senko and Prezent, and the fact that their presentation was combined with political asseveration and personal abuse, must have made these beliefs almost impossible to criticize, let alone to rebut. At their root, though, was a flat denial that genes and chromosomes are agents of heredity, or indeed that heredity is mediated through specific physical instruments so distinct from the organism as a whole; some nature-philosophers in Germany had already hinted that heredity was to be conceived as a property of every element of the organism, "a general internal property of living matter."

Genetics is not an easy subject, and it may be that much of Lysenkoism grew out of Lysenko's intense resentment at being confronted by a subject too difficult for him to grasp. He repeatedly displayed a fundamental lack of understanding of genetics. His failure to grasp its methodology is revealed by his taunt, "Genes cannot be seen under the microscope." (Of course, a geneticist's belief in the reality of genes and chromosomes does not in any way turn on being able to *see* them: genetics would not today be very different if microscopes had never been invented.) A comparable naiveté is the belief that mutation is too often deleterious to make it conceivable that mutant genes should be the raw material of evolutionary change. Again, because of the mystical significance he attached to hybridization, Lysenko, while denying that hybrid vigor might be the outcome of intervarietal crosses between two self-fertilizing lines of plants, was led to propound the almost comical absurdity that hybrid vigor might result from intravarietal crosses (between members of the same pure line).

Lysenkoism clearly could not be reconciled with the genetic tradition of Mendel, Weismann, Bateson, Johannsen, and the school of Morgan; matters, it was thought, would come to a head at the International Congress of Genetics that was to be held in Moscow in August 1937. However, the situation in Russia had deteriorated so markedly that the congress was twice postponed (until 1939, when it was held in Edinburgh). Premier Josef Stalin in the spring of 1937 publicly called for renewed efforts to liquidate "Trotskyites and other double-dealers." This provided a suitable cue for the journal known as *Under the Banner of Marxism* to organize a discussion in which the "philosopher" M. B. Mitin accused Vavilov of "menshevizing idealism" and allied him with the Trotsky-Bukharin-Pashukanis gang and with enemies of the people generally. At the same time, other individual scientists were singled out for obloquy and the Institute of Plant Breeding was denounced as a hotbed of reaction.

Under these circumstances it is not surprising that the president of the Standing Committee of International Congresses of Genetics, Otto Mohr of Oslo, received an official letter postponing the congress from August 1937 to August 1938 and denouncing as lies rumors about the arrest of Vavilov and other prominent Russian geneticists. ("Never believe a rumor until it is officially denied" is an old journalists' adage.) Against Vavilov a faction grew up in his own institute under the leadership of the head of the subtropical department, G. N. Shlykov. Vavilov defended himself from their scurrilous attacks by scientific reasoning and on one occasion reminded his abusive audience that Friedrich Engels in his *Dialectics of Nature* had gone so far as to describe Sir Isaac Newton as nothing less than an "inductive ass."

As we have seen, opposition to Vavilov was intensified by the exceedingly high regard with which he was viewed by the outside world. He had in fact been elected president of the International Congress of Genetics in Edinburgh in 1939, a meeting he was not allowed to attend. In August 1940 he was arrested by the NKVD. After his protracted trial with its foregone conclusion, Vavilov was placed in a windowless underground cell at the Saratov Prison, where he died, at least in part from malnutrition, after a few months.

In February 1965 Lysenko too was discredited. After a denunciation deliberately more temperate than those he had himself so often authored, he was dismissed as director of the Institute of Genetics.

There is no question of right or wrong about Lysenkoism: it is wrong. Lysenko emerges as the most significant figure in the Soviet science of the twentieth century. His influence was malignant, and his most spectacular accomplishment was to bring genetics in Russia to a standstill, thereby denying the world the impressive contribution its leading biologists might have made, and indeed had already begun to make.

MALARIA

A relapsing fever caused by a parasitic protozoon of the genus *Plasmodium,* malaria is the worst infectious illness affecting mankind. It is certainly the world's major killer disease, "the world" being taken in the ecological sense—that is, not as standing merely for London and the home counties and the eastern seaboard of the United States. Connoisseurs of sauce-bottle etymology will be pleased to learn that "malaria" derives by elision from *mala aria,* bad air, because of the widespread belief that the disease is caused by toxic effluvia exhaled by marshy or soggy land.

Malaria is or has been endemic in West Africa, in the Mediterranean basin, and in the southern United States (probably as a post-Columbian introduction). Although it has all the marks of an un-English disease, malaria was at one time endemic in the Essex marshes, and cases were being admitted to St. Thomas' Hospital little more than a hundred years ago.

The length of the fever cycle in malaria, whether every three days (tertian) or every four days (quartan), varies with the reproductive cycle of the parasite in the human body. The gene S that transforms normal hemoglobin, A, into the variant hemoglobin S, is a reliable ecological marker of places where malaria is or has been endemic. As explained in other entries (POLYMORPHISM; SICKLE-CELL ANEMIA), S confers in the heterozygous form some degree of protection against malignant malaria.

Among the names most highly revered in the study of malaria are those of Alphonse Laveran (1845–1922), who discovered the protozoan parasite causing malaria, and Battista and Bastianelli, who gave evidence that the anopheline mosquito was the principal vector of malaria—an idea corroborated by Ronald Ross's discovery of the parasites in the mosquito's gut.

It has been known since 1700 that quinine, an alkaloid extracted from the bark of the cinchona tree, has antimalarial properties; today, as the drug Proguanil (Paludrin), it is widely used as a preventive. For the long term, the World Health Organization takes the view that the anopheline vector must be eradicated by insecticides and by an attack on the most vulnerable phase of its life cycle (in which the mosquito larva hangs from the surface film of standing water). Detergents, being highly surface active, can sink the mosquitoes, and a film of paraffin can interfere with larval respiration. For many reasons these procedures have not yet been as successful as had been hoped: mosquitoes rapidly develop a resistance to insecticides, so that the treatment of standing water has only a temporary effect, besides being costly and administratively difficult.

In spite of present difficulties, malaria will be eradicated. The demographic consequences will be enormous. So many alarming population projections refer to the year 2000 that some people tend to believe that overpopulation is not yet an embarrassment. This view is mistaken: it has been appropriately said that the opening bars of the Dies Irae are sounding right now.

MAN'S PLACE IN NATURE

This phrase of Thomas Henry Huxley may be taken in alternative ways: it may connote either what man's place in nature *is*, biologically speaking—the sense Huxley had in mind—or it may connote what man's place in nature *ought to be* (as in "a woman's place is in the home"). Both possible meanings deserve some consideration.

Aristotle, who must surely have known of the Barbary "ape" (a macaque monkey from Gibraltar or Morocco), recognized the anatomic similarity of some animals to men; the differences between monkeys and men were substantial enough, however, to make assimilation of this concept uncomfortable. Perhaps the most profound cultural shock experienced by human beings until the days of Darwin was the discovery of manlike apes in Africa and Southeast Asia during those major voyages of discovery of the fifteenth and sixteenth centuries that made the world known to Europe. "Orangoutang" was Malaysian for "man of the woods"; and in due course the great systematist Carolus Linnaeus (1707–1778) labeled this creature *Homo silvestris orangoutang*. The broad classification of man as an animal is agreed to have been finally established by Julien Offroy de la Mettrie (1709–1751) in *L'homme Machine* (Leiden, 1748); Linnaeus classified man with the manlike apes as *Anthropomorpha*. Nicholas Tulp (1593–1674), the anatomist of Rembrandt's famous painting, was the first to dissect and describe a chimpanzee, though it is well known that the earliest anatomic comparison of a monkey, a "pigmie" (a young chimpanzee), and man was the work—highly commended by Huxley—of a London physician, Edward Tyson, in 1699.

On the grounds of the interfertility of human beings of all kinds, the philosopher Immanuel Kant (1724–1804) and the naturalist Georges Louis Leclerc, le Compte de Buffon (1707–1788), believed that human beings formed a single species. This was also the view of Darwin and Huxley, although Ernst Haeckel (1834–1919) and St. Vincent entertained the idea that there might be several human species.

The evidence relating to man's position in the world that Thomas Henry Huxley found most convincing was that deriving from the application of von Baer's Law. "It is only quite in the later stages in development that the human being exhibits marked differences from the young ape," Huxley remarked, "a circumstance that considered alone appears to me sufficient to place beyond all doubt the structural unity of man with ... the apes." Today there is no disagreement about the broader classification of man; in descending order of generality, human beings are vertebrate animals belonging to the order Primates, suborder Simiidae or Anthropoidea, and members of the family Hominidae and the genus *Homo*. Still the last word has not been said on the most judicious appellation for the species man. *Homo sapiens,* the term most widely used, was a good choice considering man's powers of ratiocination, but its principal disadvantage—literary rather than taxonomic—is the ease with which it lends itself to tiresome cynicisms about the contrast between man's titular wisdom and his practical lack of it. Other appellations have been *Homo faber*, which draws special attention to man's tool-making capabilities, and *Homo docens,* which embodies the view that the distinctive characteristic of man is to teach and learn, generation by generation, all that belongs to the human cultural heritage.

So much for the first of the two senses which we read into "man's place in nature." We may now ask what place man *ought* to occupy with regard to his status—which it seems to us idle to deny as the highest product of evolution. "Ought to be" makes it clear that we have to do with moral matters lying outside science narrowly considered. The one word that summarizes what man's relationship ought to be to the rest of creation is not monarch, not overlord, not even master—but rather *trustee*. It is a lesson that should be taught from elementary school on, that human beings have an obligation to posterity to keep the system of nature going—to pass on their own inheritance in a state not worse than that in which they received it, and preferably better. The human beings that constitute any one generation should be dismayed at the thought that they will be dismissed by their descendants as vandals, guilty of wanton destruction, genocide, and activities of the same moral stature as piracy.

It is the philosophy of commercial advantage, and above all the dismissal of those who repudiate it as do-gooders, "wets," and sentimentalists, that does the mischief—that condones the murder of elephants and whales. Our entry on ANIMALS AND HUMAN OBLIGATIONS examines this disgrace in some detail.

There are a number of structural and behavioral differences between men and the manlike apes which Huxley correctly disregarded as trivial; among them are the fact that the human hand is purely prehensile, no longer used in locomotion, and the human foot is fully plantigrade: the sole of the foot lies flat on the ground. Chimps and gorillas, in contrast, walk on the outer margins of their feet and in locomotion use the knuckles of their hands rather than the palm.

In general, human behavior is less instinctual—less genetically programmatic—than that of lower animals, and this is the sense of Maurice Merleau-Ponty's aphorism that it is in the nature of man not to have a nature. More is said of all these matters under the entry EXOGENETIC HEREDITY.

That men and apes are cousins was a lesson that had to be taught, but now it is time to remember that the differences between apes and men are much more meaningful than their common characteristics. Curiously, because Darwin did not do so, Huxley expressly excludes the higher qualities of man such as pity, affection, and tenderness as irrelevant to his argument; yet by far the most important of the qualities that distinguishes men from animals is the ability to exercise moral judgment. No: Huxley's case, as true today as when he first put it forward, was that "if any process of physical causation can be discovered by which the genera and families of ordinary animals have been produced, that process of causation is amply sufficient to account for the origin of man."

Many biologists will probably feel that Charles Darwin should have the last word. In the closing paragraph of *The Descent of Man* he wrote: "We must acknowledge, as it seems to me, that man with all his noble qualities, with sympathy which feels for the most debased, with benevolence which extends not only to other men but to the humblest living creature, with his god-like intellect which has penetrated into the movements and constitution of the solar system—with all these exalted powers—Man still bears in his bodily frame the indelible stamp of his lowly origin."

MEIOSIS

A form of cell division characteristic of gamete formation in organisms that reproduce sexually, meiosis consists in principle of two successive mitotic divisions accompanied by only one division of the nucleus. A gamete mother cell, like the ordinary somatic cells of the body, is diploid—that is, chromosomes are present in pairs, the chromosomes of each pair having been derived one from each parent. The effect of meiosis is that the gametes resulting from it contain not the diploid number of chromosomes but half that number only, each chromosome corresponding to one only of the paired chromosomes of the gamete mother cell: the haploid condition. The segregation of the two chromosomes in meiosis is at random and so is their reunion at fertilization—two random events, segregation and recombination, that are responsible for the probabilistic character of Mendelian inheritance.

MEME

This is Richard Dawkins' term for that which is transmitted in exogenetic heredity, in other words, the cultural unit passed on from generation to generation. On grounds equally of priority and etymology, it should rightly have been called a mneme. Priority is Ewald Hering's, and the root is the Greek *mnēmos,* a remembrance or record. If the word and the concept it stands for are found useful, they will stick; if not, they will not.

METABOLISM

The molecular transformations and interactions that provide for an organism's growth and being constitute its metabolism. That part of metabolism which has to do with the building up of larger molecules from smaller ones is usually referred to as anabolism, in distinction to the breaking down of larger into smaller molecules that occurs in digestion and in the processes that liberate energy, which is known as catabolism.

MIMICRY

In a general sense, any adaptation of structure, behavior, constitution, or coloration that causes one organism to resemble another or to resemble its background may be referred to as mimicry, but the term is most frequently used to describe adaptive resemblances between two or more organisms for the advantage of one or both.

In Batesian mimicry, the mimic comes to resemble an unpalatable, aggressive, or "distasteful" species. It is not surprising, therefore, that bees, wasps, and hornets are the subject of mimicry by insects of many other orders, though such mimicry may also begin by being no more radical than imitation of the conspicuous aposematic (warning) coloration of many aggressive or distasteful organisms—itself an adaptation that may be thought of as a biologically economic device to warn off predators. In Müllerian mimicry, a number of obnoxious species come to resemble one another, to the common advantage of all members of the mimicry circle. Mimicry may be behavioral as well as structural: the larvae of flukes, for example, sometimes move in such a way as to resemble the natural prey of the birds that are their intermediate hosts. Some schistosomes secure protection from the immunologic defenses of their hosts, imitating their hosts antigenically so that they are not recognized as "nonself."

The imitation by one organism of another organism or of its environment is often of such extraordinary exactitude as to make nonbiologists exclaim incredulously at the idea that natural selection can be responsible. In defense of orthodox theory, however, it should be pointed out that in mimicry we are normally dealing with situations embodying very high selection pressures, so that even fractional, and to human judgment imperceptible, variations in the direction of resemblance to the model can be quickly incorporated into the genome.

MISSING LINKS

If reptiles evolve on the one hand into birds and on the other hand into mammals, animals should exist (or have existed) which are intermediate between them and are (or more often are modern representatives of) a bridge between the two. Such animals are vulgarly but vividly known as missing links, even after they have been found. One example is archaeopteryx; among others are the monotremes, which include the duck-billed platypus and the spiny anteaters of Australasia. The concept of missing links has always been accepted as a matter of course by those who take for granted the Platonic notion of a great chain of being.

MITOCHONDRIA

Rod-shaped or spherical organelles found in the cells of eukaryote animals and plants, mitochondria are found also in egg cells and in spermatozoa. By this means they pass from generation to generation, for they are biogenetic, self-reproducing structures that are not synthesized de novo. Although most DNA is confined to the nucleus, mitochondria are unusual in containing their own DNA. Their physiological importance is that they are the principal seats of cellular respiration, but their greatest interest to philosophically minded biologists lies in the extreme likelihood that mitochondria arose in evolution as bacterial symbionts of higher organisms—a hypothesis compatible with their appearance, mobility, biogenetic origin, DNA content, and response to a variety of inhibitors of protein synthesis.

MITOSIS

This is the name given to the process in cell division by which the cell nucleus splits into two genetically identical daughter nuclei through the partition, one to each daughter cell, of the two chromosomes into which each parental chromosome divides before mitosis begins. In development the clone of cells that gives rise to the adult organism descends by mitosis from the fertilized egg, so that each cell in the body has the same complement of genes—though of course different genes may express themselves in different cells. Biologists are confident of the genetic equality of cell divisions in the embryo because the two daughter cells into which the zygote divides—or may divide after its first cleavage— are able each to give rise to a whole organism (though in some organisms each may give rise to a half-organism only). In the entry on CLONES, moreover, we point out that whole plants may be raised vegetatively from culture of a single cell. Finally, it is possible experimentally to insert an ordinary somatic nucleus from embryonic tissue—the nucleus of an embryonic liver cell, for example—into a zygote whose own nucleus has been extruded, after which, unless the experiment has foundered technically, normal development will ensue. Contrast MEIOSIS, described in a separate entry.

MOLECULAR BIOLOGY

Molecular biology is a success story of modern biology in the same way that comparative anatomy, and comparative zoology generally, were in the quarter-century following publication of *The Origin of Species*. The aim of molecular biology is to interpret biological structures and performances in explicitly molecular terms.

A shrewd and discerning historian of modern biology has pointed out that molecular biology embodies two major streams of thought: that which has to do with structure and that which has to do with genetic information. The structural stream may be said to have originated in W. T. Astbury's pioneer attempts to unravel the essentially crystalline orderliness of proteinous structures such as blood clot fibers, silk, and keratin (the structural protein of hairs, of the outermost layer of the skin, and of nails, claws, and horn). Among the greatest triumphs of structural molecular biology is elucidation of the three-dimensional structures of the muscle protein myoglobin and of adenovirus-12.

The informational stream of molecular biology constitutes *molecular genetics,* which forms an almost uninterrupted lineage of thought that can be traced back to an article published in 1928 by a medical officer of the British Ministry of Health. Fred Griffith, who was interested in the strange ups and downs of virulence in bacteria, called his paper "The Significance of Pneumococcal Types." We summarize below the part of it that relates to the future growth of molecular genetics.

Pneumococci grown upon culture plates form smoothly outlined (S) or "rough" (R) colonies, depending on whether or not the pneumococci are ensheathed in a polysaccharide capsule. This capsule may be any of several different chemical types, distinguished one from another by immunologic methods. Griffith found that living R cells derived from pneumococci of one type could be transformed into S variants of *another* type by inoculating living R cells into mice together with relatively large doses of heat-killed S cells of the type into which the R cells were to be transformed.

Such a phenomenon could not but fire the imagination of biologists. Research on the subject was prosecuted vigorously, especially at Columbia University and the Rockefeller Institute for Medical Research, where it was soon found that the agent transforming R into S cells could be rendered into a cell-free extract of which the active principle would pass through a bacteria-retaining filter. Deproteinized and polysaccharide-free extracts were still active.

It was a memorable moment in the history of biology when Oswald Avery and his colleagues at the Rockefeller Institute announced that in all probability the transforming agent was a "highly polymerized and viscous form of desoxyribonucleate"— that is, was an undenatured DNA. In the excited discussion that followed this remarkable discovery it was soon recognized that the transforming agent was, in effect, a solubilized gene and that nucleic acids therefore were in all probability the vectors of genetic information. Many biologists took offense at this interpretation, for they had grown up with and taken for granted the concept that so important a molecular function could be performed only by protein. Nevertheless, the Avery hypothesis resisted all attempts to unseat it—attempts that were very properly as searching and vigorous as so significant a discovery deserved. The hypothesis came through unscathed and no substantive modification was called for.

Collateral evidence from many different quarters confirmed the interpretation that nucleic acids are the vectors of genetic information: nucleic acid was found to be an integral constituent of viruses, and when bacteria were invaded by bacteriophages (bacterial viruses) it was the nucleic acid–bearing moiety of the virus that entered the bacterial cell. As to ribonucleic acid, the work of histochemists such as Torbjörn Caspersson and Jean Brachet had made very clear the association between protein synthesis and the presence of RNA.

The most illuminating single discovery in the whole story of molecular genetics was that whereby the two primary streams of molecular biology, structural and informational, merged into one. This consummation came about in 1953, when Francis Crick and James Watson elucidated the crystalline structure of DNA, showing how the structure lent itself not merely to the primordial act of replication—the formation of two molecules of DNA where there had been only one before—but also to the transmittal of genetic information (see the definition of GENETIC CODE). The mainstream of molecular genetics soon became confluent with biochemistry in the series of experiments that showed how the information contained in DNA is copied onto a "messenger" RNA and then in turn through the mediation of a "transfer" RNA is mapped into the structure of a protein in terms of its constituent amino acids. Many separate groups of research workers using different experimental systems made possible our present understanding of the process; and although no complete list could be compiled without detailed historical and bibliographic research, the names that rapidly became household words among contemporary biologists—in addition to those we have already mentioned—are those of Sidney Brenner, Matthew Meselson, Marshall Nirenberg, George Matthaei, Jacques Monod, François Jacob, Severo Ochoa, and Paul Zamecnic.

It could only be remarkably poor judgment or envious resentment that would deny molecular biology's claim to be the source of the most brilliant discoveries in biology since the days of Darwin and the great post-Darwinian generation of comparative anatomists. Among the many lessons that can be learned from the growth of molecular biology is the magnitude of the contributions made by physicists and chemists—a fact that wholly undermines the trite complaint that today's scientists are so specialized that they cannot even communicate with their fellow scientists, let alone with the general public. How can this view be sustained in the face of such clear evidence of the giant contributions of physicists to general biology?

The joining of the two mainstreams of molecular biology and the fusion of both with biochemistry constitutes the "molecular revolution" in biology. So successfully have molecular biologists fulfilled their function that the need—except perhaps for pedagogic purposes—for a separate and distinct discipline of molecular biology may now be questioned. It is no longer a delicate plant that needs a special soil and anxious attention: it is now the pride of the garden.

At the risk of introducing a sour note to what has otherwise been a panegyric, it is only proper to inject a word of caution. There is always a risk with revolutions of this magnitude that molecular biology may become an abuse, may usurp the attention of students and research workers at the expense of other highly valuable pursuits. There is indeed a danger that the sequencing of a protein, the more bizarre the better, may in itself be regarded as a feat deserving the most rapid preferment and the highest academic honors.

MOLLUSCA

This large and extremely varied phylum numbering tens of thousands of species includes slugs and snails, mussels, oysters, and scallops, and the relatively primitive chitons ("Venus's archpreservers"). Mollusks are soft-bodied and unsegmented animals, generally speaking, and their shells where present (in slugs they are not) are external. Nevertheless, the most advanced mollusks (the most advanced invertebrate animals, in fact) have internal shells. These are octopuses and squids, comprehensively "cephalopods," which have a relatively highly developed nervous system. They are far more mobile than other mollusks and enjoy a behavioral versatility that has made them admirable subjects for research correlating behavior with the nervous system. Squids also have astonishingly large ("giant") nerve fibers, which have been invaluable for biophysical investigation of the conduction of nerve impulses.

MONOTREMES

Primitive mammals that represent a form intermediate between reptiles and mammals, monotremes must be closely similar to animals that marked the evolutionary transition from mammallike reptiles to true mammals.

Some of the characteristics of monotremes are reptilelike, others mammallike, and others still sui generis. The most striking of the reptilian characteristics is the retention of the habit of laying shelled eggs; in addition, the rectum and the urinary and genital exits open to the exterior through a common pathway, the *cloaca*. Obviously mammalian characteristics are the possession of mammary glands similar in fine structure to those of higher mammals and—three characteristics that go together—the possession of fur, a diaphragm separating the thoracic from the abdominal cavity, and warm-bloodedness. (The diaphragm makes possible a respiratory rate higher than that of reptiles, and this in turn makes possible the higher metabolic rate associated with warm-bloodedness. Temperature regulation is not yet, however, as efficient as it is in mammals.) Another mammallike characteristic is the reduction of the lower jaw to a single bone, the dentary, and the liberation of other jaw bones characteristic of reptiles to transmit vibrations from the eardrum to the inner ear.

Among the distinctive characteristics of monotremes may be counted the absence of teeth in adults and the possession by the male of a poison gland communicating with a grooved spur on the ankle. The brain, by and large, is mammalian in structure but the *corpus callosum,* the nervous connection between the two cerebral hemispheres, is missing. Several species of ant-eating monotremes are known, and they have the modifications associated with ant eating in higher mammals.

It goes without saying that the characteristics of monotremes would be quite inexplicable except on the basis that they represent an evolutionary stage intermediate between reptiles and mammals. (See the entries on EVIDENCE OF EVOLUTION and RECAPITULATION.) The best-known monotremes are the duck-billed platypus (*Ornithorhynchus*) and the spiny anteater (*Echidna*).

MULES

Although horses and donkeys are allocated to different species (*Equus caballus* and *E. asinus* respectively), they are interfertile: the progeny of a male donkey (jackass) with a mare is referred to as a mule and of a stallion with a female donkey (jenny) as a hinny. Mules of either kind can be credited with several millennia of service to mankind, though their usefulness has been diminished by the mechanization of agriculture. The proverbial obstinacy of donkeys and mules need not be attributed to anything more deep-seated than their use among peoples habitually callous to animals and indifferent to their welfare.

MYASTHENIA GRAVIS

A chronic and progressive paralysis of the voluntary muscles, myasthenia gravis is accompanied by inflammatory destruction of the thymus gland and eventually ends in total incapacity of the afflicted individual. As with many other autoimmune diseases, the natural history is marked by alternating episodes of acerbation and remission.

The similarity—first remarked by Dennis Denny-Brown—between the symptoms of this disease and of poisoning by curare led to the introduction of anticholinesterases to bring symptomatic relief. Present-day clinical treatment is founded upon the recognition that myasthenia is autoimmune in character. Immunity reaction is directed against the protein characteristic of the acetylcholine receptor at the junction between a motor nerve fiber and the skeletal muscle it innervates. Immunosuppressive drugs are used, often accompanied by plasmaphoresis—the withdrawal of blood followed by the washing of the blood corpuscles and their resuspension and reinjection in bland plasma substitute lacking the antibodies that cause the mischief.

The involvement of the thymus, itself sometimes removed, is explained by tissue culture of thymus tissue: in the culture are embryonic precursors of muscle cells which can differentiate into contractile cells containing the acetylcholine receptors that are targets of the autoimmune attack in myasthenia.

NATURAL SELECTION

The principal agency of evolution according to both classical Darwinism and the prevailing orthodoxy—neo-Darwinism—is natural selection. We know from his correspondence with Asa Gray that Darwin was well aware of the animistic flavor of the phrase, for Nature does not pick and choose, but he excused himself on the grounds that the term could be avoided only by the use of lengthy and tiresome periphrases.

Any population of which the elements are (a) self-reproducing and (b) subject to mortality—a population such as that formed of individual animals or of genes—may be the subject of natural selection. In other words, one or another element may enjoy a net reproductive advantage over the remainder. This in turn means only that their reproductive rate and mortality are such that the "favored" elements increase their proportional representation in the population as time goes on—or, in an equivalent formulation, that the elements naturally selected can be said to enjoy greater Darwinian fitness than the others.

The power of natural selection to change the characteristics of a population through differential fertility or mortality is not merely a theorem deduced from first principles. Natural selection can be, and has been, shown to work in nature, for example by increasing the representation in the population of better mimics, or of organisms generally better able to cope with the hostility of the physical environment.

The *subject* of natural selection (that which is or is not selected) may be an individual organism, a gene, or even a community of organisms. The entries on GROUP SELECTION and NEO-DARWIN-ISM and the definition of POPULATION GENETICS are all relevant.

NATURE AND NURTURE

The antithetical use of the terms "nature" and "nurture" probably derives from Sir Francis Galton's *English Men of Science: Their Nature and Nurture* (London, 1874), but in *The Tempest,* act 4, scene 1, Shakespeare puts into Prospero's mouth the description of Caliban as "a devil on whose nature nurture can never stick." Robert Nisbet in his essay on "Anomie" in *Prejudices: A Philosophical Dictionary* (Cambridge, Massachusetts, 1982), feels that the antithesis is foreshadowed by an antagonism dating from the fifth century B.C. between *physis* (having much the connotation of "nature") and *nomos* (meaning a pasture for cattle). The usage is in any event long established and well understood.

In the modern application of these terms, an organism's nature is that which it is by reason of its genetic makeup. Its nurture is the sum total of the factors that have entered into its environment and its upbringing. Laymen are apt to think of the distinction in competitive terms and have been wont to ask themselves what fractions nature and nurture respectively contribute to a certain character difference—say, a difference of intelligence. We can only proffer a general rule about attempts to attach exact percentages to the two contributions: the more confident and dogmatic the attempt, the more likely it is to be wrong—or, more often, wrongheaded. Indeed, the estimated contribution made by nature may vary from 0 to 100 percent according to environment, that is, according to the contribution of nurture—and vice versa. This point is made in our entry on PHENYLKETONURIA, the inborn error of metabolism which is responsible for approximately 1 percent of all cases of mental deficiency.

Phenylketonuria as we recognize it in an ordinary environment is caused by the inheritance from both parents of a certain recessive gene whose conjunction affects the individual in which they are conjoined in such a way as to make it impossible to metabolize the common dietary amino acid phenylalanine. Afflicted individuals are raised on a diet with greatly reduced phenylalanine content, and under these conditions the inability to metabolize phenylalanine is not the menace it otherwise would be. Consider now two boundary situations: in a real world abundant in phenylalanine, the disease may be rightly regarded as of genetic origin; but in a microcosm in which all the individual members are homozygous recessives, phenylketonuria would be confidently described as a disease of environmental, that is of nurtural, origin. It would be said that one of the amino acids liberated by the digestion of most proteins is an extremely toxic substance, *phenylalanine,* exposure to which must be reduced to a minimum.

These two may be regarded as limiting cases illustrating a theorem of general validity which declares that the contribution of nature to any character difference is a function of nurture and of nurture is a function of nature, a theorem of which there are an infinitude of illustrations. For example, in many crustaceans the degree of pigmentation of the eye depends both upon the genetic makeup and upon the temperature at which development proceeds: it is possible to fix upon a temperature at which the differences between having white, black, red, or dusky eyes present themselves as wholly genetic in origin, and it is possible to choose genotypes in which the color of the eyes is determined almost wholly by environmental influences. Exactly the same argument in principle applies to the number of facets in the compound eye of insects: this too is under genetic control but also influenced by temperature.

If in these simple and straightforward cases it is seen to be quite impossible to attach exact figures to the contributions of nature and nurture to character differences, must this not apply a fortiori to those weary attempts by IQ psychologists to demonstrate that intelligence is 75 to 80 percent genetically controlled and only influenced to the tune of 20 to 25 percent by education and nurture? Such pronouncements have had a profound influence upon politicians and through them upon legislation—sometimes, especially in California, leading to the enactment of cruel, stupid, and inefficacious laws to counteract hazards as illusory as the remedies proposed for them. We are here thinking especially of the "danger" that the supposedly riotous fertility of the feebleminded will soon populate the world with imbeciles who will eventually supplant those of sound mind.

The argument that has been outlined above, that the contribution of nature is a function of nurture and of nurture a function of nature, is not in principle a very difficult argument, especially if one guides one's thoughts by the boundary examples chosen for illustration. Nevertheless, to understand it requires a certain strength of comprehension which seems in the past to have been beyond the capacity of many psychologists, high though their IQ scores may have been.

There is a strong political coloration to the inclination to attribute inequalities of intellectual performance to nature rather than to nurture. For if the poor and needy have become so by reason of their genes, there is nothing very much we can do about it: their poverty and inadequacy are not of our doing, and we have no moral compulsion to take social steps to remedy their condition. If, on the other hand, like dedicated modern Marxists we categorically deny that there are inborn differences among human capabilities so that a human being is only what his environment and upbringing can make of him, then in a just society it is an obligation upon the state to provide for the education and upbringing of its citizens.

It is the faith of scientists—touchingly naive though it would be thought to be—that the only remedy for these uncertainties, and for the injustice and maladministration to which they give rise, is the discovery and promulgation of the truth. Naive, perhaps, because one of the consequences of partisanship is to deafen disputants to arguments that go contrary to their constitutional beliefs. The attempt to discover and to promulgate the truth is nevertheless an obligation upon scientists, one that must be persevered in no matter what the rebuffs—for otherwise what is the point of being a scientist? The alternative is for scientists to content themselves with being the mere handymen and artisans of a complex machine-based culture.

NEO-DARWINISM

When Charles Darwin formulated the theory that Alfred Russel Wallace generously described as Darwinism, neither he nor Darwin had any correct notion about the nature of inheritance, though both realized that natural selection could not be an agent of evolution unless the variants upon which it was declared to act were the products of *heritable* variations.

The new geneticists—Thomas Hunt Morgan and William

Bateson and R. C. Punnett—were very well aware of the inadequacies of the evolutionary theory considered as a theorem in natural history. Consciousness of these imperfections led Bateson to remark in his great *Mendel's Principles of Heredity* (Cambridge, 1909) that so deep-seated was the belief that Darwin had done it all that in the years following Darwin the study of anything that concerned the problem of species was "marked by the apathy characteristic of an age of faith."

In spite of this evidence of dissatisfaction it was not until the 1920s and 1930s that Sewall Wright, J. B. S. Haldane, and R. A. Fisher propounded the methodology and principal theorems of a Darwinism refounded upon Mendelian genetics—in short, a neo-Darwinism.

In this new Darwinism the subject of selection—that which is selected—is the gene, and the concept of gene frequency is especially important. The charge of unreality so often brought against gene frequency has been dispelled by the Hardy-Weinberg Theorem, which makes it possible to translate statements about gene frequencies into statements about the frequencies of genotypes in the population.

The fundamental principle of neo-Darwinian theory—that which should form the baseline to all our thinking—is the conservation of genetic variance: gene frequency in an interbreeding population remains constant from generation to generation except insofar as some impressed "force" is brought to bear on the population. A force properly described has magnitude and direction; the magnitude of any force that affects gene frequency is measured by the rate of change of gene frequency, and its direction is given by the sign or sense of the gene substitutions in progress (for instance, whether an allele A' is substituted for a gene A or vice versa).

Several agents may cause a change of gene frequency and thus be agents of evolution. The first is drift, that is, errors of random sampling in the parentage of a succeeding generation as a result of which the gene frequency fluctuates and the proportions of genotypes in the next generation do not come out according to the book. When chance has it that departure from randomness works in the same direction in several succeeding generations, the systematic change of gene frequency is known as drift (see the definition of GENETIC DRIFT). We cannot gainsay that drift occurs, but the magnitude of its contribution to evolutionary change is still a matter of debate. Adherents of the notion attach particular importance to drift as it occurs in small populations.

A second agency that may change gene frequencies systematically is the pressure of one current mutation—the repeated substitution, for example, of mutant allele A' for gene A. However, mutation is so infrequent an event that Darwinists do not look upon mutation pressure as an important agency in evolution.

By far the most significant agency is natural selection, Darwin's own enduring contribution to biology, which has not diminished in professional esteem even though laymen and amateurs often disapprove of the notion.

Among the misconceptions about neo-Darwinism none is more widely prevalent than the one which presents it as a declaration that what Nature selects is mutations—or more correctly, mutant genes or mutant organisms. This is a travesty of neo-Darwinism, and we take the opportunity to reaffirm our proposition that mutations do *not* arise in response to an organism's needs, nor do they, except by accident, gratify them. Candidature for evolution is provided by the virtually infinite range of variants thrown up by the genetic recombination that is made possible by genetic segregation, the random reunion of gametes in the sexual process, and by crossing over. This prodigious combinatorial variety, with the recombinational elements further enriched from time to time by mutation, proffers to natural selection a repertoire comprising trillions of candidates—of possible organisms—any one of which might contribute significantly to a solution of the problem of remaining alive in a predominantly hostile environment.

Such is the neo-Darwinian theory of evolution by natural selection. Darwinism itself has faults, of course: it cannot be wholly acquitted of the charge of explanatory glibness and the reproach that it does not account very convincingly for evolutionary progression—that is, for the adoption of solutions that are more complicated than called for by the bare requirements of subsistence and reproductive solvency. The objection most frequently made, however, is one which cannot be sustained: that it is *inconceivable* that evolutionary adaptations with all their finesse and refinement of detail could possibly have come about by such a mechanism. The "inconceivability" complaint is a criticism of ourselves rather than of neo-Darwinism. We cannot but remember that the distinguished English respiratory physiologist John Scott Haldane declared it "inconceivable" that there should exist a chemical compound having exactly the properties deoxyribonucleic acid is now known to possess, as an engram of heredity, a system of instructions for the execution of development.

Neo-Darwinism has its own faults, of course. Like Darwinism, it has a measure of glibness. Moreover, it is difficult to think of any experiment or set of observations that could possibly falsify it: one such test would be to establish a heritable difference by artificial selection among a population of highly inbred, indeed homozygous, mice. Thus if Samuel Butler and William McDougall had been right in their surmises about the heritability of behavior, neo-Darwinism would be completely undermined by improving learning ability by selection of inbred mice throughout an experiment lasting many generations on the same diet. Even if any biologist were willing to undertake such an experiment, no grant-giving body known to science would be willing to fund it.

NEOTENY

The relative rates of development of the several parts of the body, including the gonads, are not invariant. One of the most interesting (and from an evolutionary point of view one of the most important) departures from a set tempo and order of differentiation is that whereby organisms become sexually mature at a relatively juvenile stage. What is in effect an adult therefore retains juvenile, even larval, characteristics in adult life. The general term for the phenomenon, coined by Gavin de Beer, is pedomorphosis.

Neoteny, the authenticity of which cannot be questioned, is best exemplified by the Mexican axolotl, the larval form of an American salamander of the genus *Ambystoma* into which it may be induced to metamorphose. Walter Garstang was one of the first to derive the higher chordates from the "tadpole" larvae of sea squirts by neoteny. The inherent plausibility of the theory is witnessed by the fact that one entire class of the sea squirt phylum, the so-called *Larvacea*, is composed of clearly neotenous sea squirt larvae.

Human beings, it is thought, owe some of their characteristics to being neotenous in relation to the great apes. Ludwig Bolk in 1926 introduced the term *fetalization* to describe the changes that caused human beings to retain some of the characteristics of fetal apes: the relatively enormous brain, the position of the foramen magnum and the rectangular set of the skull in relation to the backbone, the hairless body, the light color of the skin, and certain other characteristics. With his usual inventiveness, Aldous Huxley in *After Many a Summer* imagined a state of affairs in which extreme prolongation of life made it possible for a human being to metamorphose into an adult ape.

Considered as an evolutionary stratagem, neoteny provides for relatively rapid and radical evolutionary transformation, providing what A. C. Hardy referred to as "an escape from specialization."

NERVE GASES

A nerve gas is a volatile liquid, often an organophosphorus compound, and the vapor issuing from it effectively paralyzes its victims by its action as an anticholinesterase. The nerve gases thought most highly of by those who have a mind to use them are colorless and nonirritant and therefore extremely insidious in their action. They have some application as pesticides.

NERVE IMPULSES

A nerve is essentially a bundle of nerve fibers proceeding in the same direction. As a rule, it connects the central nervous system, brain, spinal column, and so forth with motor end organs and, conversely, connects sensory organs with the central nervous system. Such bundles of fibers have a delicate connective-tissue framework and are supplied by blood vessels but not by lymphatics. A nerve *within* the central nervous system that connects one part of it with another is referred to as a tract; an example is the "pyramidal" or "corticospinal" tract, which conveys nerve impulses from the cerebral cortex down the spinal cord.

Peripheral nerves may be motor or sensory or mixed. From the way in which the eye arises in development, it is clear that the so-called optic nerve is not really a nerve in the sense in which the sciatic nerve may be so described: it is a brain tract. The cell bodies of the sensory neurones that run in this tract lie in the retina itself.

Nervous communication is transacted by nerve impulses. A nerve impulse is a propagated change of state in the membrane of the nerve fiber; nothing material passes down the length of the fiber, any more than it does in a line of dominoes so arranged that knocking one over knocks down its neighbor and so on. This too is a propagated change of state: what moves is a *signal,* not a material substance. A nerve fiber can conduct impulses in either direction. The interface between two neurones, the *synapse,* polarizes the impulse, which can cross the synapse one way only. Signals in the nervous system work on the FM (frequency modulation) as opposed to the AM (amplitude modulation) system because trains of nerve impulses differ one from another in their frequency only: the amplitude is fixed and does not die away as the distance from the point of stimulation increases. However, if by a narcosis or some equivalent means the amplitude is caused to drop over some stretch of nerve, it picks up again to its original value when the signal has passed the affected stretch. This is evidence that nerve impulse is self-energized, that is, powered by the nerve itself.

The nerve impulse has been described by one of its principal students as "the universal currency of the nervous system." Indeed, the fact that most of the business of the nervous system including the brain is transacted by electrical signals is one of the respects in which brain and nervous system are computerlike.

NUCLEIC ACIDS

Proteins, carbohydrates, and fats have a huge variety of structural and metabolic functions to perform in the body, but the functions of nucleic acids are comparatively simple: they are the vectors of information for development and heredity. Thus its nucleic acid moiety is that part of the chromosome which conveys the hereditary message, and it is another nucleic acid that relays this message to the synthetic machinery of the cell, causing a protein to be assembled in a particular way, with its constituent amino acids lined up in a specific order.

A nucleic acid is a huge elongated molecule of a chemically repetitive (polymeric) structure of which the backbone is formed by linkages between phosphate and 5-carbon sugar molecules. These are one or the other of two kinds of molecules that give their names to the main categories of nucleic acid: *ribose* in ribonucleic acid (RNA) and *deoxyribose* in deoxyribonucleic acid (DNA). The nucleic acid of the chromosomes is DNA, which is found also outside the nucleus—in, for example, mitochondria. Attached to the phosphate backbone are nitrogen-containing organic bases that are purines or pyrimidines. The tectonic unit of a nucleic acid— the monomer of which the polymer is compounded—is a *nucleotide,* composed of phosphate, sugar molecule, and organic base; it is the order of the organic bases down the length of the molecule that encodes the genetic information or genetic message that passes from one generation to the next. In the synthesis of a protein, which requires the assembly of constituent amino acids in a particular order, the ribonucleic acids (which are of different functional species) act as go-betweens. First, the specific order of nucleotides in the DNA molecule is mapped onto a strand of messenger RNA. Upon the structural basis of a minute cytoplasmic particle known as a ribosome, this messenger RNA marshals yet another (soluble) species of RNA known as transfer RNA, each molecule attached to a specific amino acid. By this procedure information contained in the original DNA molecule is mapped into the structure of a protein. By convention, the transaction between DNA and RNA is known as transcription, while the RNA is said to be translated into the structure of protein.

It is widely believed that the researches which uncovered the structure of nucleic acids, explained how they fulfill a genetic purpose, and demonstrated how they convey information in cellular metabolism, are among the most brilliant and illuminating scientific advances of the twentieth century. With this interpretation we concur. It was a long business, though, for when nucleic acids were first recognized as chemical entities, the methods of extraction were so violent, often involving heating in an alkali or the like, that the crystalline structure of the nucleic acid was quite destroyed. The researches that finally established the complicity of nucleic acids in genetic information transfer are summarized in the entry on MOLECULAR BIOLOGY. The initiative was in some sense a conceptual one. Only when the gene was conceived as a *message* did there arise a special incentive to associate these gene messages with the particular crystalline structure of a macro-molecule such as nucleic acid. It is the linear differentiation of the molecule that provides for information transfer, but replication is made possible by the fact that the helical molecule is double stranded, the two strands being complementary in such a way that either half may be made the basis from which a complementary half can be assembled.

Order in Biology

If we were to be confronted with an unfamiliar object, whether of microscopic size or large enough to see, then ask if it were a living organism or the remains of one, we should almost certainly base our answer on the degree and kind of order in the object being inspected. For order is everywhere evident: in an organism's definite shape and distinctive symmetries; in the crystalline tidiness of chromosomes, viruses, and many of the constituents of cells; in the giant polymeric molecules that are so abundant a constituent of living organisms; and in the products of the activity of organisms such as honeycombs, spiderwebs, and the various kinds of "houses" or nests which organisms make for themselves. Order permeates biology through and through.

A system displays order when its constituent parts are not randomly assembled but manifest only one or a few of the possible configurations of their elements in space and time. So considered, any given configuration may be thought of as an improbable one, the degree of improbability being clearly related to the number of possible configurations. This reflection gives intuitively a sense of the connection between order, disorder, and probability. No generalization in physical science is more firmly established than that which declares that the direction of flow of events in the universe is always from less probable to more probable—always, that is, toward randomness and "mixed-upness"—toward the equipartition in the universe of that which began by being separated and unequally divided.

With this, as Lord Kelvin was first to discern, goes a progressive loss of the ability to use the heat in the universe for performance of mechanical work. Although the total amount of heat energy in the universe remains constant, it is "irrevocably lost to man" and therefore "wasted," though not *"annihilated,"* in thermodynamic transactions—for the performance of mechanical work by heat rests upon the existence of inequalities in the system that are progressively done away with in a vast and universal leveling.

In Ludwig Bolzmann's formulation the relationship between entropy S and order is given by the expression $S = k \log W$, where k is the universal Bolzmann constant and W the probability of the prevailing configuration of the system. Where the probability of a given configuration is unity, there is no possibility of alternative configurations. Under the circumstances, the entropy of the system is of course zero (remembering that $\log 1 = 0$). Such is the case of a perfectly pure crystal at $0°K$ (absolute zero), where heat motion is at a standstill.

The relationship between information and order is made intuitively clear by the observation that information embodies, expresses, and often specifies order. For nearly all information is imparted in the form of a linear configuration of symbols, the specificity of which depends upon the way in which these symbols are ordered. Consider, for example, the anagrams *time, item, mite,* and *emit*—four different words that are so by reason of the order of the alphabetic symbols of which they are composed. An ordinary message like that contained in a telegram would cease to convey information if its elements were randomly mixed—as in telegrams they sometimes are. Clearly, information and disorder are antithetical, and information accordingly can be defined and quantitatively measured as negative entropy. A chromosome or an architect's drawing is said to contain information because either may be conceived as a scheme of instructions for imposing or generating order. Erwin Schrödinger (1887-1961) was one of the early physicists who made significant contributions to biology. In his illuminating little essay *What Is Life?* (Cambridge, 1944) Schrödinger was the first to propose that the phenomena of heredity turn upon a coded *message* transmitted from one generation to the next, the message being embodied in a chromosome, considered as an aperiodic crystal, in which the structure is such as to be relatively insulated from the disordering effects of heat motion. See the entry on NUCLEIC ACIDS.

Do living things circumvent or flout the Second Law of Thermodynamics? Rudolf Clausius' famous dictum that entropy of a closed system tends always to a maximum value is as true of living organisms as it is of the physical world. Still, it is very natural that the pervasiveness and ubiquity of order in biological systems should have created the impression that evasion of the Second Law is a distinctive and crucially important characteristic of living things. This viewpoint is mistaken, however, and grows out of failure to realize that the law applies only to *closed* systems—systems in which there is no external traffic in matter or energy. In a closed system the increase of order in one place is matched by an increase of disorder elsewhere—as it is, for example, in a domestic refrigerator, the cooling of which must be paid for by the heat generated in the pumps that compress the refrigerant gas and circulate it so that it can "excrete" its excess heat into the environment. A living organism, said Schrödinger, "drinks orderliness from the environment"; he was referring to the continual degradation of small molecules in the course of metabolism, which pays the thermodynamic bill for the generation of complexity, normally of a higher order. Nevertheless, it is fair to say, as Aharon Katchalsky more than once has: "Life is a constant struggle against the tendency to produce entropy. Since there is no escaping the entropic doom imposed under the second law of thermodynamics, living organisms choose the lesser evil—they produce entropy at a minimal rate by maintaining a steady state."

The production of a symmetrical molecule, or of a structure such as a spiderweb with its symmetrical features, leads to a greater loss of entropy than the production of an asymmetrical structure. The thermodynamic cost of symmetry, however, is minutely small. The entropy loss involved in spinning a spiderweb would not, because of the symmetry, be significantly more than the loss already associated with the burning of sugar molecules to compensate for the physical exertion of spinning (unless, of course, the spider did more work to spin a symmetrical web than to spin an untidy or haphazard one).

Robin Valentine once called our attention to a striking example of the degree to which changes of order are entirely inapparent. He cites the formation of tobacco mosaic virus (TMV) from its subunits. TMV is a highly ordered crystal, and its formation must lead to a loss of entropy; entropy elsewhere must therefore increase. So we should expect a rise in temperature of the solvent—but in fact the solvent cools. This, we were told, is because the surfaces of the virus subunits are hydrophobic and therefore induce order in the water by which they are surrounded. When the units aggregate, their hydrophobic surfaces are inturned, causing the solvent to become more disorganized; this outweighs the order created in the virus. In this case, then, the Second Law actually favors the creation of a well-ordered virus.

James Clerk Maxwell (1839–1871) envisaged a situation in which the Second Law of Thermodynamics was circumvented by a being with intelligence and a very high degree of perceptiveness. Let us suppose, said Maxwell, that a vessel is divided into two portions A and B by a partition in which there is a small hole. An observer able to see the individual molecules opens and closes this hole so as to allow the swifter molecules to pass from A to B and the slower ones to pass from B to A. He will thus without expenditure of work raise the temperature of B and lower that of A, in contradiction to the Second Law of Thermodynamics. Maxwell's "demon," as he came to be called, highlights the notion that information is negative entropy.

ORDERS OF MAGNITUDE

In biology very small quantities are come upon more often than very large ones, and the accompanying table will serve as a memorandum of the terminology used for orders of magnitude.

Name of fraction	Proportion of a meter	Order of magnitude (power of 10)
Meter	Unity	0
Decimeter	Tenth	−1
Centimeter	Hundreth	−2
Millimeter	Thousandth	−3
Micrometer	Millionth	−6
Nanometer	Billionth	−9
Picometer	Trillionth	−12

We need to remember too that a designation such as 10^{-3} stands for $1/10^3$, that is, one thousandth, and that by convention $10^0 = 1$.

A corresponding terminology applies in gravimetry, in which the unit is one gram: thus milligram, microgram (μ), nanogram, and so on. In volumetry, the unit of volume is one liter: one milliliter is approximately one cubic centimeter and one microliter is about one cubic millimeter.

Latin prefixes are used for quantities less than unity and Greek prefixes for quantities greater than unity; thus *deca-*, *hecto-*, *kilo-*, and *mega-* signify the multiples 10, 100, 1,000, and 1,000,000. In professional writing extremely large or extremely small figures are almost always represented as powers of ten, positive or negative as the case may be. The orders of magnitude of the quantities above, then, are (+) 1, 2, 3, and 6 respectively.

ORIGIN OF LIFE

Scientists who believe neither in miracles nor in spontaneous generation have naturally felt themselves under obligation to give an account of how living systems may have arisen under the conditions prevailing on the surface of a primeval earth from inorganic precursors, the prior existence of which was taken complacently for granted. So probing the origin of life has grown into a minor research industry which some practitioners and lookers-on regard as important, exciting, and absorbing. It is only fair to say that there are others who regard the whole subject as a painful bore. They are in good company: from a letter to Sir Joseph Hooker, we take it that Charles Darwin was one of them, and though he does not explicitly say so, Erwin Chargaff is probably another.

Research into the origin of life is not of vital importance in any sense that the advancement of learning depends upon its successful completion. Experiments to date show that the exposure of carbon dioxide and water to forty million electron-volt helium ions from a sixty-inch cyclotron may lead to the formation of minute quantities of formaldehyde, which would certainly disappear very quickly under the oxidizing conditions that currently prevail on the earth (though not perhaps so rapidly in the reducing atmosphere Aleksandr Oparin has postulated for the primeval earth). In other experiments in which Harold Urey has played an important part, mixtures of water, methane, and ammonia exposed to electric discharges, ultraviolet radiation, and some other sources of energy may form formaldehyde, hydrogen cyanide, and by secondary reactions even amino acids and purines, essential ingredients of protein and nucleic acid molecules respectively.

From time to time it has been claimed that meteorites contain organic molecules such as hydrocarbons and even amino acids, but the possibilities of contamination are so great that these claims have been greeted with extreme caution.

Needless to say, these findings represent only a partial solution to the problem—and not the more important part, either. Beyond mere assembly of the raw materials for biological molecules, an adequate theory must postulate some means by which they can recruit more molecules and enlarge their own substance, a process that will surely turn upon the formation of a membrane and a primitively cell-like structure.

The present experiments do not go very far, but they make one point of considerable philosophical importance: the origin of life from inorganic precursor substances may be in the very highest degree unlikely, but the phenomenon is not *inconceivable*. However dim it may be, a first conception of how life might have come about is a necessary condition for future progress; without it, there can be none.

ORNITHOLOGY

Part science and part natural history, ornithology owes its tremendous popularity as a science and a scientific recreation to a simple circumstance that is sometimes forgotten when we weigh the eligibility of different kinds of animals for informal study: birds are day animals where others, just as engaging in their way—mice, for instance—are creatures of the night. Apart from their visibility (something very much in their favor), the beauty of birds and their many appealing habits have won for them a huge fan following that has more than once tempted impatient and ill-informed laboratory biologists to dismiss ornithology as so much bird-watching—an absurdly unjust judgment that ignores the fact that ornithology has made a number of profoundly important contributions to general biology.

Professional biologists are not hard put to reel off a list of dis-

tinctive contributions that ornithology has made to general biological science. They will cite, for example, Ernst Mayr's investigations of evolution and the mechanisms of speciation, or M. R.
Irwin's work on the immunologic performance of species of *Colum-
bidae* (pigeons and doves) and of hybrids between them. Then
again, David Lack's studies on mortality in wild populations of
birds are the most illuminating life tables of wild animals and of
animal demography generally. Most important of all, perhaps, is
the contribution that ornithology has made to the study of animal
behavior, beginning with Julian Huxley's classic study of the
courtship habits of the great crested grebe and followed by Niko
Tinbergen's studies on the behavior of herring gulls. Nor, in the
compilation of such a credit sheet, should it be forgotten that because birds lay eggs in abundance, which are easy to incubate and
hatch, their study has contributed enormously to our understanding of developmental processes and of parental care.

PARTHENOGENESIS

A variant of sexual reproduction, parthenogenesis is the process by which an egg develops without fertilization by sperm. The offspring may sometimes be haploid, but normally they have the diploid number of chromosomes. Parthenogenesis is the regular rule in some animals such as rotifers and plant lice (aphids).

The phenomenon is not known to occur in human beings, and young ladies should be warned that there is a test whereby a claim that a child is of virgin birth may be verified: under normal circumstances a skin graft transplanted from a child to its mother will be rejected by the mother in the usual fashion of allografts, but a skin graft from a child of parthenogenetic origin will not be rejected because it can contain no gene, and therefore no antigen, not also present in the mother. When this test was applied to a woman whose claim to virgin motherhood was being sponsored by an English tabloid newspaper, the graft was duly rejected—a circumstance which led the newspaper to infer that the test must be unsound.

"Artificial parthenogenesis" is the initiation of cell division in an unfertilized egg, say of a newt or a sea urchin, by artificial stimulus such as pricking or osmotic shock.

PHENYLKETONURIA

Phenylketonuria (or PKU, as it is known) is a rare but extremely grave inborn metabolic disorder, an uncomplicated example of recessive inheritance, for it is the biochemical expression of a gene that must be inherited from *both* parents if it is to exercise an outwardly discernible effect. PKU illustrates very clearly the manifold bodily effects of single gene differences. The primary biochemical lesion is a deficiency of the enzyme in the liver that normally causes the breakdown of *phenylalanine*, an amino acid that is a ubiquitous constituent of the normal human diet. The immediate consequences of this enzymic deficiency are an accumulation of phenylalanine in the body and the excretion of phenylpyruvic acid in the urine. The most serious consequence of this accumulation of phenylalanine is imperfect development of the brain, leading to mental deficiency: indeed, it has been estimated that about 1 percent of the mentally retarded are unrecognized victims of phenylketonuria. Among the other abnormalities resulting from this primary biochemical lesion is interference with the production of the coloring matter melanin; the result in untreated victims is a characteristic pallor of skin, hair, and eyes. Further secondary consequences of untreated PKU include disturbances of temperament such as overactivity, sleeplessness, anxiety, and sometimes epileptiform seizures. The state of affairs in which the causal pedigree of a variety of ostensibly unrelated symptoms can be traced to a single gene substitution is known as *pleiotropy* of gene action and is the rule rather than the exception.

Phenylketonuria can now be recognized very early by biochemical tests used as a matter of routine in the countries of northern Europe and North America. Treatment consists in a reduction of dietary phenylalanine until the anatomic differentiation of the brain is complete. This is a procedure of the kind classified by Joshua Lederberg as "euphenic"—that is, an improvement of the environment sufficient to diminish the effect of a deleterious gene.

A *eugenic* treatment of PKU might take one of two forms. A normal gene might be substituted for its deleterious recessive form, a procedure not yet known to be feasible, although analogous procedures can be executed in microorganisms. The second possibility involves preventing the conjunction of the two recessive genes which when present together are responsible for the biochemical lesion. This is an entirely feasible procedure whenever the carriers of the deleterious recessive gene (heterozygotes) can be identified. The carriers will not normally be gravely afflicted, or even perceptibly at a disadvantage, because the adverse effects of the recessive gene are masked by the action of its normal dominant partner.

Nearly all victims of PKU are the offspring of parents who are both carriers, for the genetics of the situation is such that when two carriers bear children, one quarter will be normal, one quarter will be afflicted, and a half will be carriers like their parents. The frequency of PKU would plummet in one generation if its carriers could be discouraged from marrying each other and would accept a cordial recommendation to marry someone else instead, a someone else sure to be a carrier of a harmful recessive gene but not, if humanly avoidable, the *same* recessive gene.

To many, many people—especially those who have appointed themselves guardians of civil liberty and the conscience of us all—the idea that anyone should not marry whom he or she chooses is seen as a gross invasion of personal liberty, a wanton withdrawal of natural rights. This is a most questionable argument, for who has conferred upon human beings the right to bring genetically crippled children into the world? and how many would insist on standing on their rights to the extent of knowingly doing so? Although it does not affect the principle of the situation, it is highly relevant to the practical use of preventive eugenics that the threat to human liberty entailed by this eugenic procedure is quantitatively very minor. A recessive disease with a frequency of, say, one in ten thousand births is the victim of a deleterious recessive gene with a frequency (p) of about 1/100 in the population. According to the Hardy-Weinberg Theorem, where the gene frequency is 1:100 the frequency of the afflicted homozygotes, p^2, is 1:10,000, and the frequency of the carriers to whom we are referring, the heterozygotes, will be $2p = 1:50$. The outcome of this is that the frequency of marriages between carriers will be 1:2,500 (1/50 × 1/50). On the average, therefore, only one in 2,500 particular marriages would then be contraindicated through the carriage of this particular gene, and each disappointed partner would in theory have forty-nine others to choose from.

These figures, of course, paint too bright a picture because we may be obliged to grapple with not one but several such deleterious recessive diseases, and partners judged eligible with respect to one recessive disease may conceivably flunk when other genes are taken into consideration. Moreover, the quotation of frequencies that relate to the population as a whole is not likely to do much to comfort those who suffer the distress associated with being warned against a marriage which on other grounds had been deemed suitable—unless of course the partners, with almost inhuman equanimity, can derive an equivalent satisfaction from having avoided the risk of bringing a biochemically crippled child into the world.

A caution is necessary. The restoration of victims of PKU to something approximating a normal life by the use of diets low in phenylalanine, and the discouragement of marriages between carriers of the offending gene are measures strongly dysgenic in tendency and would make the PKU problem worse in succeeding generations. Their effect is virtually to cancel the work of natural selection in eliminating the PKU genes from the population. The constant reintroduction into the population of the offending genes through mutation is no longer counterbalanced; for it is only through the deaths of those that possess it that a gene can be reduced in frequency. Under either of the regimens of containment recommended, the bad gene is not removed from the population and there is nothing to prevent its frequency from slowly increasing through the recurrent process of mutation.

PILTDOWN SKULL

Between 1908 and 1912 a number of supposedly human remains were discovered by Charles Dawson and several colleagues in a gravel pit at Piltdown, in Sussex, England. Associated with them were a number of artifacts—worked tools of flint and bone, and a number of fossil remains dating the find as early Paleocene.

These remains were dignified by the specific name *Eoanthropus dawsoni* ("dawn man," Sir Arthur Smith Woodward's *Earliest Englishman*). The finds were unintelligibly anomalous, the braincase being domed and unmistakably human and the lower jaw being that of an ape. The anomaly was not resolved until two British physical anthropologists, with the moral support of a famous anatomist, declared the skull to be a fraud: to simulate great age the bones had been steeped in salts of chromium, and to simulate wear the teeth had been artificially filed down. Later it became clear that the fossil remains accompanying the skull had been deliberately planted to give an air of antiquity to the discovery.

There is no conclusive evidence incriminating any one man, although suspicion has fallen on all the principals in the discovery, not excluding Pierre Teilhard de Chardin, one of Dawson's assistants. Teilhard in later years was the author of a work, *The Human Phenomenon,* of the kind seventeenth-century scientists would have dismissed contemptuously as a "philosophic romance." Unless a confession by the real culprit comes to light, the true answer is not likely ever to be known. All the same, the term "hoax," widely used to describe the episode, is unduly euphemistic: this was a carefully executed crime carried out with the intention of deceiving.

PINEAL BODY

A dorsal outgrowth of the *diencephalon,* which lies between the end brain and the thalamus, the pineal organ had an eyelike character in lower vertebrates, or was in any event light sensitive. In present-day mammals it is thought to be an endocrine gland affecting pigmentation.

The pineal body is more evident in man than in lower animals, and its prominence in embryonic development is something that could not have been known to Descartes. Perhaps this is why he proposed the pineal body as the seat of the soul (he may have been influenced by the fact that he did not believe animals had souls). Descartes's "soul" was not anything that could be contained in a material body, as the gallbladder contains bile; rather it was an animating principle—"a wind, a flame, or an ether"—in which resided the power of thought, especially of imagination.

POIESIS

A suffix familiar in biology, *-poiesis* derives from the Greek root *poiesis,* standing for making or creating (it is the root also of the English words "poesy" and "poetry"). The activities referred to as erythropoiesis, lymphopoiesis, and the like refer to the generation of red blood corpuscles, lymphocytes, and so on. As to the poetry, it is clear that when Percy Bysshe Shelley in his *Defence of Poetry* said that "poetry comprehends all science," he was implying that the act of creation involved in the composition of poetry was in its essence akin to the imaginative exploit that is the origin of all scientific ideas.

POPULATION GROWTH AND CONTROL

The growth in numbers of biological populations illustrates an important and perhaps a defining characteristic of biological growth: *that which is formed by growth is itself capable of further growing.* This is the distinguishing mark of growth by continuous compound interest, for if the starting number is regarded as capital and the additional members of the population formed by growth as interest, then clearly the interest is added to capital and begins to earn interest on its own behalf. Growth of this kind is often referred to as exponential or logarithmic.

The sense of this nomenclature is immediately obvious if we consider the simple case of a population of organisms, each member of which divides into two in each unit of time; the size of the population at each unit of time will be as shown in the accompanying table. The logarithms in the last line are necessarily the same as the exponents in the second last line, for these logarithms are taken to the base 2, the logarithm being the power to which 2 must be raised in order to give the number of organisms in the population.

Time units	0	1	2	3	4	5	10
Number in population	1	2	4	8	16	32	1,024
Exponential formulation	2^0	2^1	2^2	2^3	2^4	2^5	2^{10}
Logarithm to base 2	0	1	2	3	4	5	10

It can be seen by inspection that if one draws a graph relating the logarithm of population size to elapsed time, the points will fall along a straight line, whereas if we had instead plotted the arithmetic number against time, the points would form curves sweeping rapidly upward as exponential curves do.

No real population could grow in exponential style for long: in real life the growth of populations is always restricted by the action of one or more *density-dependent* factors, usually having to do with living space, inadequacy of nutrition, accumulation of toxic waste products, stress associated with overcrowding, and so on. It is thus a mere figure of speech to say that human populations can ever become so numerous that they will occupy the entire land surface of the globe shoulder to shoulder, perhaps even two deep; it is no figure of speech, though, to say that the density-dependent factors most likely to act upon human populations will be famine and pestilence of unimaginable severity.

The past hundred years have been marked by the partial annulment of some of the most important density-dependent factors that act on human populations. Thanks to the exertions of medical men and sanitary engineers, we have achieved a measure of *death control* that has not been counterbalanced, unluckily, by the degree of *birth control* necessary to keep the human population within reasonable bounds. The population explosion is the well-known and frightening outcome, and a striking example of the unpredictability of many of the malefactions consequent upon the growth of science.

The exigence of the population problem prompted many politicians and prominent personages to call for the stabilization of human populations. This is a feasible-sounding enterprise, and we should all be very fortunate if it were as easy as going to the moon. A population can become stable in numbers only when appropriately matched regimens of fertility and mortality have been at work long enough for the population to achieve a stable age distribution, whereupon it will reproduce itself faithfully to that pattern. No human population has ever achieved this stable state, and in our opinion there is never likely to be a time in which we are not worried alternatively by fears of a precipitous decline in the human population and fears that the population will grow to a politically and economically unmanageable size.

At one time the idea gained ground, especially in the United States, that if all couples would restrict themselves to having two children, stability of numbers would be achieved. "This proposal," it has been remarked, "has about it that air of common-sensical rightness which is almost invariably a symptom of some aberration of reasoning." The aberration is that no weight is given to mortality or infertility, important though they both are. An average completed family size of two would be a sure recipe for the catastrophic decline in human numbers which demographers feared in the 1930s.

Consider the difficulties that would be faced by a hypothetical Minister for the Control of Population: in the first place, he would have to provide for the discovery of better methods of birth control in the face of the natural conspiracy in favor of fecundity (to which we owe our own presence here on earth), and these methods would have to be made available to populations who do not know how and may not want to learn how to use them: that alone is a formidable political, religious, and educational enterprise. In a free society, one in which the members cannot be required either to reproduce themselves or to abstain from doing so, the Minister of Population would have to work as best he could through fiscal and educational means—by bribing, coaxing, cajoling, or pleading with people to have a greater or lesser number of children as the prevailing situation required.

PRIMATES

This is a term introduced by Linnaeus to signify monkeys, apes, and human beings. Professional zoologists often pronounce the word in three syllables, Latin style, though there is no serious likelihood of confusing the connotation of Linnaeus' term with that which describes a synod of the principal officers of Anglican and Episcopalian churches. Linnaeus allotted primates the first rank among animals; in second place he put mammals other than primates and these, appropriately enough, he designated *secundates*. All other animals were of the third rank, *tertiates*. (We learned of Linnaeus' distinction between primates, secundates, and tertiates from J. Z. Young's *Life of Vertebrates,* 2nd ed. [Oxford, 1962].)

Primates are an order of the class Mammalia: they are typically long-limbed arboreal animals with many adaptations associated with the arboreal habit and movement by brachiation. Among these are opposable big toes with thumblike functions, dominance of the visual faculty, and binocular vision with forward-facing eyes—a characteristic that endows animals for the first time with something that is recognizably a face.

The primates descended from an insectivoran stock, probably from animals much like the tree shrews. Curiously, though, the faces of primates are the basis of a taxonomic distinction: primates with a narrow nasal septum are Catarhines and comprise the Old World monkeys, anthropoid apes, and man. Those with a broad nasal septum and a flattish face are the New World monkeys, Platyrhines.

It was surely their similarity to man and to the animals closest to man (the Anthropomorpha) that induced Linnaeus to designate primates as the first among animals. This must in turn lead us to designate human beings as the first among primates, a rank of which only bad judgment or an adolescent style of cynicism could deprive them.

Among the most manlike characteristics of the great apes (especially chimpanzees) are, as Jane Goodall has observed, the beginnings of exogenetic heredity: the transfer of information from generation to generation through nongenetic channels. This is the very characteristic to which human beings owe their position as first among primates.

PROTEINS

In the entry dealing specifically with them, we said of NUCLEIC ACIDS that their essential function was to act as vectors of genetic information in communication between generation and generation and between nucleus and cytoplasm.

The baroque profusion of different kinds of proteins and the multiplicity of their functions are such that no such tidy story can be told: the contractile element of muscle fiber proteins is responsible for all muscular movement, whether in locomotion or in the working of heart and guts; as enzymes, proteins are involved in almost every metabolic transaction of the body. They are the principal structural constituent of cells; they are responsible for the clotting of blood and lymph; and as connective-tissue fibers (collagen, elastin), they prevent the Organism-as-a-Whole from falling apart.

Proteins are huge polymeric molecules, "polymeric" because they are built upon the repetition of unit building blocks, "monomers." These are the amino acids, of which there are some twenty different kinds, each one coded for by a triplet of *nucleotides,* in the nucleic acid molecule that specifies their assembly in a particular order. The primary structure of proteins is given by the linear order of the amino acids along the carbon chain into which a protein molecule might in theory be unraveled. A protein's secondary structure specifies the pattern of branching, and the tertiary structure the pattern of its folding into a three-dimensional structure. The complete three-dimensional structure of myoglobin and of the protein capsule of adenovirus-12 are now known—both splendid feats of analysis.

Proteins differ sufficiently in size to make centrifugation of their solutions an effective method of separating them. Bearing as they do a net positive or negative charge, they migrate slowly in electric fields with the effect that electrophoresis may also be used for separation.

Proteins, typically, are soluble in water but the protein keratin that is the principal ingredient of hair, nail, and claw and the outermost layer of the skin is not. Water-soluble proteins cease to be so when denatured by heat or by the action upon them of the salts of heavy metals. Familiar examples of denatured proteins are the white of a hard-boiled egg and the skin that forms on hot milk when left standing. A process akin to denaturation occurs when the water-soluble protein *fibrinogen* is converted by enzymes into the insoluble fiber *fibrin,* which is the structural basis of a blood or lymph clot.

Proteins, moreover, are characteristically antigens. Their immunologic properties may be used to distinguish them in a much more sensitive and discriminating way than is possible by any merely physicochemical procedure.

Proteins are nutritious and as skeletal muscle—the form most highly prized as a food—they are normally associated with a quantity of fat that quickly dispels the illusion that "butcher's meat," far from being fattening, is positively thinning in character.

PROTOPLASM

More than fifty years ago the biologist and mathematical logician Joseph Woodger wrote: "Biologists certainly speak as if there were in nature a stuff to which the term 'protoplasm' is given.

When we come to look into the question there does not seem to be any justification for this belief" (*Biological Principles* [London, 1929], p. 293). Unexpectedly, Hans Driesch, the vitalist, had anticipated the modern practical usage: "Protoplasm is a mere name for what is not the nucleus."

We have pointed out elsewhere that in real life the word is now only used as a general term for the cell sap, but in the old days when it was believed there was a substance to which the term "protoplasm" might properly be applied, there was a good deal of discussion about its fine structure. There was a granular theory, a reticular theory, and indeed as many theories as there were fixatives for hardening tissues in order to make microscopic preparations of them. For such fixatives almost invariably contain coagulants of protein, and it was the different protein precipitation patterns that gave rise to the different theories of protoplasmic structure. Thomas Henry Huxley had no doubt about the physical reality of protoplasm: "I have translated the term protoplasm which is the scientific name of the substance . . . for the words 'physical basis of life'" (*Fortnightly Review,* 1 Feb. 1869). Protoplasm was a kind of nature-philosophical ether: just as the ether was thought to permeate all material structures in the world and act as the medium of propagation of electromagnetic waves, so the living substance protoplasm was thought to permeate otherwise inanimate structures and endow them with vivacity.

The influence upon Ernst Haeckel's mind of the idea of protoplasm, abetted by that of a great chain of being, led him to suspect the existence of very lowly organisms consisting of "naked protoplasm"—of *Urschleim,* the primitive slime—only. Such organisms he professed to have discovered, and he called them *Monera.* Not to be outdone by what was at that time a rather competitive science, Huxley in the *Quarterly Journal of Microscopical Science* described a specimen of monera from a deep North Atlantic dredging, and out of respect for the learned zoologist of Jena he named it *Bathybius haeckeli.* Today the word "protoplasm" is falling into disuse, a fate that would certainly not have overtaken it if it had served any useful purpose, even as a first approximation to reality.

The demise of the protoplasm concept resulted from the slow realization that life processes have a *structural* rather than a colloidal basis: even Frederick Gowland Hopkins' famous pronounce-

ment that life is a manifestation of a complex dynamic equilibrium in a polyphasic system no longer carries much weight. The
earliest attempts at X-ray crystallography of bodily constituents
combined with electron microscopy soon made it clear that the
orderliness of living cells and their products is essentially a solid or
crystalline orderliness, and that the little organelles in the cell are
minute "solid" bodies looking as if it was only their size that prevented our picking them up and handling them. Thus the speci-
ficities of metabolism in time and space—of why metabolic events
happen here and not there, now and not then—has a structural
basis, something very far removed from primordial slime.

PROTOZOA

That moiety of the lay public which knows that the formula
H_2O designates a molecule of water knows also that "the
amoeba" is an elementary and rather lowly living organism—"a
mere speck of protoplasm" is the popular cliché. Amoeba is indeed a protozoon, and it is very small indeed. Schoolchildren
anxious to make a good impression on their teachers or examiners
should guard against making remarks such as "If one pricks the
amoeba with a pin ..." This is a form of words which instantly
reveals a total lack of comprehension of how small amoebas are
and how large, relatively speaking, pins are.

Considered collectively, protozoa are adjudged to constitute a
phylum, and although they do not lend themselves to the full-
dress hierarchy of classification as it applies to many-celled animals, they may be usefully and naturally subdivided into a number of classes that are distinguished mainly by the ways in which
they move.

Reference has been made elsewhere to the quasi-philosophical
nonproblem of whether individual protozoa are to be described
as noncellular or as single cells. Nothing turns on the decision,
which embodies not much more than a nuance of meaning: those
who think of cells as mechanical or administrative subdivisions of
an organism tend to refer to protozoa as noncellular because they
are not so subdivided; on the other hand, those who think of the
cell as the tectonic unit, and of a many-celled animal as a colony

of cells enjoying a high degree of integration and division of labor, tend to refer to protozoa as single celled. Upon one subject, however, all zoologists agree: protozoa are *organisms,* with all that that implies of wholeness and degree of integration of parts—and also of self-reproduction.

One of the classes into which the protozoa are conventionally subdivided, the Sporozoa, is something of a ragbag of mainly parasitic forms which are not easy to enter under the names of other classes.

Many protozoa are parasites, and it is mainly as disease-causing agents that they obtrude themselves upon the attention of human beings. The trypanosomes that infect cattle, *Trypanosoma brucei* and *T. congolense,* are not wholly maleficent; in a sense they tend to protect the resident immune cattle population against competition from immigrants or imports from elsewhere.

Trypanosomes are responsible for two very serious human illnesses; the better known is sleeping sickness, the work of *T. gambiense* or *T. rhodesiense.* Tsetse flies are notoriously the vectors of trypanosomes, which may invade brain and central nervous system to cause extreme lethargy and sleepiness (hence *encephalitis lethargica*). In South America *T. cruzi* causes a disease of protean manifestations, Chagas' disease, which invades heart muscles and visceral ganglia. Charles Darwin was almost certainly a victim of Chagas' disease, a hypothesis indignantly repudiated by a number of physicians for a variety of ostensible reasons that conceal what we believe to be the principal motive: strong resentment that medically unqualified scientists have so far forgotten themselves as to have attempted a medical diagnosis. This may be combined with the uneasy and well-founded feeling that Darwin's treatment at the hands of his own physicians (embodying as it did the theoretically unfounded use of arsenicals and mercurous compounds, not to mention treatments such as "Dr. Gully's water cure") shows the medical profession in a rather unfavorable light. We are amused to notice that Brazilian medical opinion has eagerly accepted the hypothesis of Darwin's having been a victim of South American trypanosomiasis—as if his having suffered a disease so closely associated with the names of two famous Brazilian medical scientists (Carlos Chagas and Osvaldo Cruz) were a source of some national pride.

Trypanosomal infections sometimes respond to treatment by organic arsenicals, though these may have grievous side effects. Protozoan parasites are responsible also for malaria, but not for syphilis, since spirochetes have for some time been regarded as bacteria.

PTERODACTYLS

Reptiles twice took to flight. One line of descent—that which leads to birds—is represented by an intermediate form such as archeopteryx; the other line of descent is that represented by pterodactyls. Although *Pteranodon* had a wing span of more than twenty feet, pterodactyls generally are quite small. Their musculature, the essentials of which can be inferred from the structure of the skeleton, is such as to make it certain that they could not fly by flapping wings. The likelihood is that pterodactyls were gliders, equipped with a flight membrane supported in front by an extremely elongated fourth digit of the hand. The aerodynamics of their flight as it relates to takeoff and landing has not yet been fully worked out.

RECAPITULATION

Ernst Haeckel's Law of Recapitulation declares that in the course of its development an animal passes through developmental stages unmistakably comparable to the forms of its presumed evolutionary ancestors. His own account of the matter was fully expounded in the fifth edition of his *Evolution of Man* (trans. Joseph McCabe; London, 1906). As we have said elsewhere, the sense of the law is sometimes given by "ontogeny recapitulates phylogeny," or by the schoolboy variant that in development "an animal climbs its own family tree."

A mammal begins life as a single cell, the ovum, ostensibly a structureless blob of "protoplasm" that might be compared to a single protozoon. In the course of development all vertebrates pass through a metamorphosis which in effect converts a hollow ball of cells into a double-layered structure such as might be created by pushing in the ball at one pole. This forms a so-called gastrula and thereby establishes a firm kinship to essentially two-layered animals such as many polyps and hydroids. Eventually the embryonic axis is laid down. The embryo at this stage is a bilaterally symmetrical structure with a well-defined head bearing paired sense organs, a dorsal central nervous system, and a median dorsal backbone. This, Haeckel pointed out, was essentially the condition of the lancelet (amphioxus), universally agreed to represent an ancestral chordate form. In subsequent development vertebrates go through a fishlike stage, the expanded anterior end of the gut tube being perforated by structures looking like gill slits.

Haeckel believed that not only the soma but also the psyche obeyed the principle of recapitulation; again in *The Evolution of Man* he wrote: "The wonderful spiritual life of a human race through many thousands of years has been evolved step by step from the lowly psychic life of the lower vertebrates, and the development of every child-soul is only a brief repetition of that long and complex phylogenetic process."

One of the most important critical texts on recapitulation is Gavin de Beer's *Embryos and Ancestors* (3rd ed.; Oxford, 1958). Sigmund Freud accepted the idea uncritically: in *Civilization and Its Discontents* (ed. James Strachey; London, 1965), attempting to justify his belief that modern civilization is the victim of neurosis, he argued as follows: "If the development of civilization has such a far reaching similarity with the development of an individual should we not be justified in coming to the conclusion that in some civilizations or epochs of civilization possibly even the whole of mankind has become neurotic?"

Haeckel's Law of Recapitulation is not satisfactory, but as we point out in the entry HYPOTHESIS AND THEORY, it cannot be pronounced bluntly to be "wrong," for it embodies an element of truth significant enough that we can readily understand why it should have been propounded.

The element of truth is the same as that which is embodied in von Baer's Law, named in honor of Karl Ernst von Baer (1792–1876), the distinguished embryologist who first recognized the mammalian ovum. The farther back we go in development, the more closely the embryos of related animals are seen to resemble one another anatomically. As development proceeds, their patterns diverge and the embryos become progressively more unlike.

One can see, then, the sense of Haeckel's Law; human beings and chickens never look in the least like fish, yet there is a stage in their development when their embryos look remarkably like the embryos of fish. Von Baer's Law implies that higher animals do to some slight degree recapitulate the embryonic history—not the adult forms—of their presumed ancestors, and this is just what one would expect. For phylogeny can only be a succession of ontogenies, so that what evolves in an evolutionary sequence is a sequence of ontogenies; it may be a necessity that the developmental history that is to be modified must first be gone through, to pave the way causally for its subsequent modification. Thus if the development of the human arm or the wing of a bird is an evolutionary modification of the development of a fish's pectoral limb, it is not surprising that the development of a wing should pass through a stage similar to that of the development of a pectoral fin, cognate reasoning applying to the development of the human arm. The various phenomena grouped together as "recapitulatory" in character and pointing to an affinity between mammalian development and the development of lower vertebrates (for example, the appearance of visceral clefts in the development of higher vertebrates) seemed to Thomas Hunt Morgan the most convincing single piece of evidence in favor of the idea of evolution. T. H. Huxley also thought so.

In spite of its substantial measure of truth, therefore, the Law of Recapitulation has faded away, or rather has been assimilated into a law of greater generality—that which we associate with the name of von Baer. Just in time, too, for as Stephen Jay Gould has pointed out in his authoritative text *Ontogeny and Phylogeny* (Cambridge, Massachusetts, 1977), the idea of recapitulation has been used by some to support the intrinsic inferiority of certain races that are held to represent mere staging posts in the evolution of a higher type of human being, such as that represented by Haeckel himself.

REDUCTIONISM

Reductive analysis is the most successful research stratagem ever devised: it has been the making of science and technology. Reductionism is the belief that a whole may be represented as a function (in the mathematical sense) of its constituent parts, the functions having to do with the spatial and temporal ordering of the parts and with the precise way in which they interact.

This is a far cry from the conventional travesty that is almost compulsively propounded by certain nature-philosophers, and by genuine philosophers also, who for obscure psychological reasons resent the whole idea of analysis. Indeed, some resent the whole idea of elucidating any entity or state of affairs that would otherwise have continued to languish in a familiar and nonthreatening squalor of incomprehension. Their travesty is embodied in the declaration that reductionists, mechanists, and other philosophic evildoers represent the whole as "a *mere* sum of its constituent parts." Summation is of course a perfectly respectable functional relationship which might well obtain between parts and which in some cases might furnish a complete description: the *mass* of a whole, for example, is the sum of the masses of its constituent parts, and the osmotic pressure of a solute is a function of the number of ions and molecules in the solution. Not even a holist would maintain that a composite body had a "true" mass as opposed to a "mere" mass, with the true mass being something more than the sum of the masses of the parts. Yet something close to this absurdity was at one time thought to account for the empirical fact that the combustion of a metal, such as magnesium, leads to an increase of weight. This phenomenon was difficult to account for in terms of the phlogiston theory, according to which phlogiston should have been given off (phlogiston, the essential principle of fire, being for inorganic chemistry roughly what Hans Driesch's entelechy was for biology). Phlogiston, it was proposed, weighed less than nothing: small wonder, then, that the loss of phlogiston on the combustion of a metal led to an increase of weight.

In biology very few properties of composite systems can be represented as merely additive functions of the properties of constituent parts—and those that can be so represented are of no very great interest. The functional relationships which obtain between parts and which must be ascertained if a description is to be complete are extraordinarily complex, and their elucidation is proportionately laborious. Indeed, the whole enterprise is so difficult and tedious that it may sometimes embody too extreme an ambition. In the tenth edition of his *System of Logic* (London, 1843) John Stuart Mill wrote as follows of sociology: "The laws of the phenomena of society are and can be nothing but the laws of the actions and passions of human beings gathered together in the social state . . . Human beings in society have no properties but those which are derived from, and may be resolved into, the laws of the nature of individual men."

A simple paraphrase converts this declaration into the form that used to cause needless dismay and resentment among "organismic" biologists: "The laws of the properties and performances of living organisms are and can be nothing but the laws of performances of living cells gathered together into the state characteristic of living organisms. Living organisms have no properties but those which are derived from and may be resolved into the properties of individual cells."

The principle of reducibility obviously could be taken one or two steps lower in the hierarchy—for cells can be resolved into molecules and, *mutatis mutandis,* much the same could be said about cells as was said about organisms in relation to cells; then again, molecules might be resolved into their constituent atoms, and they in turn no doubt into elementary particles.

It should be remembered, moreover, that the relationship "reducible to" is of the kind described by logicians as transitive: thus if A can be analytically reduced to B and B to C, then A can be reduced to C. This being so, it follows that a concept belonging to the social, political, or ecological domain of discourse can be interpreted in terms of the atoms and molecules of which the material world is composed; to put it concretely, such an upper-level concept as the foreign-exchange deficit should be interpretable in terms of the properties and interactions of atoms and molecules, and so should the idea of proportional representation—yet, oddly, there is no mention of these matters in any textbook of economics or physics or political science.

It is evident that reducibility has its limits. The success of reductive analysis in biology is due mainly to the fact that the properties of organisms that interest us most are exactly those that *do* respond to reductive treatment: the system of signaling in the transfer of genetic information, or the physical dispositions that make possible the nervous coordination of the parts of the body.

By representing a composite whole as a function of its constituent parts, we are almost automatically empowered to envisage a domain of possible wholes other than that which formed the original subject of the analysis. Indeed, perhaps the simplest way to epitomize reductionism is to envisage any particular frog as one realization in a universe of Possible Frogs, any one of which *might* have become real, though in fact only one did; likewise the whole world is seen reductively as one only of a whole domain of possible worlds—not necessarily the best. This way of looking at reductive analysis makes it clear that among all conceivable ways of understanding the world, reductive analysis is the one that makes it easiest to see how, if need be, the world might be changed.

It is of course the character of the interaction among the components that fixes the character of the structure, phenomenon, or performance under consideration: thus only a very limited subclass of the possible interactions among cells permits their integration into a living organism—and likewise only a minority of the possible interactions among persons is such that an organized society or ecological community will emerge.

Before discussing *emergent* properties, we must make an important point about the empirical content of the various levels of the hierarchy into which the system of animate nature may be said to fall. In increasing order of particularity we may tabulate the hierarchy thus:

Society/community

Organisms

Organs

Cells

Cellular organelles

Macromolecules

Molecules

Atoms

Subatomic particles

A significant property of this arrangement is that the higher levels include the lower levels: societies are composed of organisms, organisms of organs, organs of cells, and so on. The sciences that deal specifically with each level, starting at the bottom with particle physics and ending with sociology or ecology, increase in empirical richness as we ascend the rungs of the hierarchy. Each may be said to comprehend the concepts and factual matter of its own subject plus the concepts and factual matter of all lower-level disciplines, with sociology/ecology necessarily being the richest of all, for everything that is true of organismic biology is true also of sociology. It is a sociological—not merely a physiological—truth that living organisms that require food exert themselves to procure it, and it is also a truth of sociology that the liver is rapidly damaged by ethyl alcohol and a whole variety of other organic solvents. It is a truth of organismic biology, no less than of chemistry, that the interaction of an acid with an alkali produces salt and water:

$$\text{NaOH} + \text{HCl} = \text{NaCl} + \text{H}_2\text{O}.$$

This is also a truth of politics, though not one that is likely to sway the electorate.

As the empirical content (the repertoire of actual or possible objects of observation) increases, so also does the conceptual content—not necessarily in predictable ways. Indeed, each level of the hierarchy has concepts peculiar to itself and not interpretable in terms of the lower levels; *foreign-exchange deficit* has already been mentioned (it belongs to the political level, and propositions relating to it make no sense at any lower level). Again, at the level of atomic physics and chemistry it is possible to understand how one atom of carbon may unite with two atoms of sulfur to form carbon bisulfide, CS_2; but CS_2 has a notoriously bad odor not interpretable at any physicochemical level, to which even the concept of smell does not belong; for smell, along with sexuality, memory, and fear, is contextually peculiar to the level of organismic biology.

Properties or phenomena that appear at higher levels of integration and are not predictable or interpretable in the lower ranks may be described as *emergent*. C. Lloyd Morgan, a champion of the notion, drew a useful distinction between resultant and emergent properties, the former being understandable and indeed predictable in terms of lower analytic levels (properties such as mass, osmotic pressure, and combustibility), whereas emergent properties cannot be predicted or interpreted by reductive analysis.

The concept of emergence is not offensive. It is unpopular only because those who have successfully employed the stratagem of reductive analysis tend to feel that the designation "emergent" implies a loss of face (a feeling quite uncalled for). The real objection to the notion of emergence is that it has no explanatory value: it marks the end of one train of thought, not the inception of a new one. We are not more deeply understanding or wiser people for resolving that the possession of mind is an emergent property, for so it clearly is: there is no psyche or rudiment of psyche in a crystal or in a molecule. Nor is there any rudiment of love, though Pierre Teilhard de Chardin maintained that if love were not present in some primordial form in atoms and molecules, it could not have made its appearance at the level of evolution represented by mankind. Thus the union of a sodium ion with a chloride ion would be construed as a premonition of that union between higher organisms which was the outcome of love—an argument defended by an English admirer of Teilhard by pointing out (and we do not deny it) that if sodium and chloride ions had not possessed a propensity to unite, love could hardly have appeared at a later stage in evolution. This defense unhappily embodies a serious confusion of thought: failure to distinguish between necessary preconditions and sufficient preconditions for a phenomenon, event, or state of affairs is an egregious logical error.

The idea of emergence plays a useful part in the biological sciences if only by giving a name to that which does not respond to reductive analysis, though it must be added that biological research of all kinds prospers proportionately as more and more of its subject matter does respond.

REFLEX

The definition of a reflex proposed by B. F. Skinner may be put as follows: in any episode of behavior where S represents a stimulus and R a response, we can speak of a reflex action when $R = f(S)$, that is, the response is a specific function of the stimulus. The reflex concept is of much wider application than in neurophysiology; thus immunologists have sometimes spoken of an "immunologic reflex" when referring to the specific relationship between the antigenic stimulus and the immunologic response it gives rise to. It has been helpful to immunologists to distinguish an afferent from an efferent arm of the immune response; the former relates to events having to do with the presentation of antigen at the response center, and the latter to events having to do with the way in which the immune response is put into effect.

A number of subordinate notions have grown up around the reflex concept; the term "reflex arc," for instance, has been used to describe the supposedly one-track pathway by which a stimulus excites in turn a sense organ, an afferent nerve, and, after making synapse in the central nervous system, an efferent nerve and a motor end organ of some kind.

Holism and nature-philosophy sharply repudiate an opinion

attributed to, but not held by, Charles Sherington or any member of his school (nor indeed by anyone except nature-philosophers inventing imaginary opponents on whom to launch attacks). This is the opinion that all reflex behavior can be resolved into the sum of individual reflex arcs, something which is of course very far from being the case. Indeed, the reason the "reflex arc" terminology went out of use was precisely because Sherington himself and his school showed very clearly how interactions in the central nervous system and the convergence of afferent pathways upon a single efferent pathway made the reflex arc a hopelessly unrealistic concept.

The term "conditioned reflex" refers to a state of affairs in which a reflex response is excited not merely by the stimulus that originally gave rise to it but by a conditioning stimulus that had repeatedly been caused to precede the original stimulus. Conditioning gives rise to the neurological equivalent of an expectation and provides also for a rudimentary form of afferent learning, that is, learning to increase the repertoire of stimuli to which a certain motor response may be given.

RHESUS FACTOR

When blood of rhesus monkeys is injected into rabbits, an antibody is formed that reacts upon the red blood corpuscles of approximately 85 percent of human beings. These are rhesus positive, the remaining 15 percent being described as rhesus negative. In keeping with the nomenclature of other blood antigens, several factors that make up the rhesus antigen are usually referred to as C, D, and E—of which the most important clinically is D. Whereas antibodies to blood antigens A and B occur naturally in some human beings, antirhesus antibodies are formed only by active immunization (for example, by transfusion into rhesus-negative individuals of blood which, though compatible for other important antigens, contains the rhesus antigen). Or rhesus antibodies may be formed when a rhesus-negative mother bears a number of children who inherit the rhesus antigen from their father. Immunization of the mother may occur during parturition by leakage of fetal blood cells into the maternal circulation. When this unhappy state of affairs occurs, maternal anti-bodies entering the rhesus-positive fetus may cause extensive destruction of red blood corpuscles, which accordingly brings about a burst of proliferative activity among blood-forming cells (*erythroblastosis fetalis*).

Sensitization is much less frequent when mother and infant are incompatible with respect to red-cell antigens of the A, B, O system—the reason being that if fetal red cells are of group A and the mother is of group B or group O, she has ready-made anti-A antibodies that will destroy fetal red cells in her circulation before they can arouse antirhesus immunity. This protective effect, it has been thought, can provide the selection pressure that keeps A, B, O polymorphism in being. The selective advantage that maintains rhesus polymorphism itself is not known unless perhaps it is a transient polymorphism that will eventually die out.

SENSE ORGANS

An instrument of the kind described by engineers as a transducer, a sensor translates one energy form or waveform into another. Some examples are the phonograph pickup and the microphone, which translate mechanical and atmospheric pressure vibrations respectively into voltage pulses. This is analogous to the action of biological sense organs, with the difference that the latter translate environmental changes of state into nerve impulses. A sense organ is specially adapted to respond to one kind of environmental change of state rather than another—to temperature, for example, or to light, pain, sound, posture (as disclosed by muscle and joint sensors), to orientation in space and angular acceleration (both the work of the ear, an equilibratory as well as an acoustic organ). Furthermore, a sense organ lowers the threshold of strength of stimulation needed to initiate a nerve impulse: sensory adaptation (a process in which a sense organ ceases to respond to a stimulus of uniform intensity) is described in the entry on ADAPTATION. It is a phenomenon that reminds us that sense organs can respond only to a *change* of state.

One of the most important generalizations that can be made about the physiology of sense organs is that embodied in Johannes Müller's Law of Specific Irritability, or Law of Specific Nervous Energy. In effect, this law states that the modality of sensation—whether of light, sound, pain, or posture—is determined not by the sense organs, which after all can only generate nerve impulses, but by the nerve itself, or rather by its central connections. Stimulation of the auditory nerve produces sensations of sound, and of the optic nerve sensations of light, by whatever means the nerves are stimulated. Much else in sensation is centrally governed: thus different cortical receptors provide for the perception of a line that is vertical and one that is tilted.

In all these respects it can be seen that sensation follows a Kantian pattern. Immanuel Kant's contention in *The Critique of Pure Reason* was that that which we perceive or become empirically aware of is determined by the character of our faculty of intuition or perception. Kant would have been delighted to know to what a degree modern sensory physiology has confirmed his insight. He went on to say, as we need not, that some knowledge can be a priori, that is, independent of all experience. Lancelot Hogben argued that this position is undermined by the existence of a sense that Kant was unaware of, namely proprioception, the joint and muscle sense upon which postural self-awareness is based; but we do not find this argument convincing.

The central tradition of British (as opposed to Continental) philosophy, especially as associated with the name of John Locke (1632–1704), is the empiricism embodied in the line *Nihil in intellectu quod non prius in sensu*—There is nothing in the mind but what entered it by way of the senses. Biology cannot sustain this belief. While we do not agree with Kant that there can be knowledge independent of all experience, the empiricist axiom fails by disregarding the existence of behavioral programs that constitute a kind of "inherited knowledge." Such is the case with a mouse's knowledge of nest building, or a bird's of its particular song. Whatever part upbringing and sensory experience play, a large part of their specificity—the property of being *this* song and not *that*—resides in the chromosomal recording (one might almost say tape recording) that is part of an organism's inheritance. Equally, there is nothing in biology to give scientific authority to Cartesian rationalism, which accepted as true that which rested upon an intense inner conviction of the self-evident truth of certain propositions, combined with an inability even to imagine any skeptical argument that could shake them. It is well known that one such proposition, avowedly the first principle of Descartes's philosophy, was *Cogito ergo sum*—I think, therefore I am.

Because all that we know of the working of the senses is based on sensory evidence, it is of no use to appeal to physiology to rescue us from the Berkeleyan idealism according to which the external world exists only by virtue of the ideas of it we form in our minds. We cannot better Samuel Johnson: "I refute it *thus*," said Johnson, kicking a large stone so violently that he rebounded from it—"thus," it has been remarked, simultaneously refuting George Berkeley and corroborating Sir Isaac Newton's Third Law of Motion ("Every action has an equal and opposite reaction").

SESSILE ANIMALS

One does not have to be a professional zoologist to call to mind without difficulty a variety of animals that are sedentary because they are permanently attached to the substratum on which they live; among them are sea anemones, hydroids and polyps generally, barnacles, sea squirts, sea lilies (crinoids), and limpets. Sessile animals share a number of structural and physiological adaptations: they show a tendency to radial symmetry and develop either tough external skeletons or thick, resistant, and virtually inedible sheaths or tunics. The power of regeneration, and the asexual reproduction that so often goes with it, is highly developed. Hermaphroditism tends to be the rule, accompanied by various devices to prevent self-fertilization; the fertilized eggs usually develop into mobile larvae, which undergo a radical metamorphosis before fixation to the substratum.

SEXUAL CYCLES

Whether estrous or menstrual, sexual cycles may be thought to have evolved in response to the selection pressures arising out of the necessity to conserve and to facilitate fertilization of the limited number of oocytes with which the female is endowed. Because male mammals manufacture sperm throughout life, there is no imperative reason to conserve it. It is not therefore surprising that males are not cyclic in their capacity or apparent inclination to mount females; female mammals, on the other hand, are receptive only at intervals that recur cyclically during the period known as estrus—from the Greek *oistros,* the gadfly that bites cattle and maddens them in a way believed in classical times to simulate the exigence of sexual desire.

Ovulation, the shedding of an oocyte, occurs just before the onset of heat. Although in New World monkeys vestiges of this cyclic receptivity linger, human beings, apes, and Old World monkeys are characterized by a menstrual cycle of a somewhat different kind. The word "menstruation" itself refers to the periodic shedding of the *endometrium,* the highly vascular internal lining of the uterus into which the fertilized egg is implanted. A number of vascular adaptations reduce to a minimum the loss of blood. Ovulation occurs halfway through the menstrual cycle, in an episode sometimes accompanied by mittelschmerz, or midpain.

The occurrence of some vaginal bleeding at the height of estrus was long a source of confusion between the estrous and menstrual cycles. However, the two are clearly distinguished by the fact that the female is receptive throughout the menstrual cycle (a circumstance which, Sir Gavin de Beer proposed, played an important part in the social evolution of the one-to-one pairing of male and female, monogamy).

SEXUALITY

The teleology of sexuality is not in dispute: it provides for genetic recombination, and in turn for evolutionary versatility, by enormously enlarging the candidature for natural selection. Compared with the rival attractions of parthenogenesis and asexual reproduction, sexuality is clearly a Good Thing—but that, of course, is not explanation enough of its having evolved.

Genetic recombination is a very ancient biological stratagem that must have existed long before the evolution of sex. Bacteria have sexuality, but some have also the power—doubtless related to the existence of a DNA repair mechanism—to "infect" each other with their indigenous DNA, as shown by the discoveries of Griffith and Avery that we have elsewhere recounted. Once established in the population, sexual dimorphism is enforced by the chromosomal mechanism described in the definition of SEX DETERMINATION.

It is difficult to think of any evolutionary process other than group selection that could account for the maintenance of sexuality in competition with ostensibly simpler reproductive mechanisms. Still, as John Maynard Smith says in *The Evolution of Sex*

(Cambridge, 1978), "even if we accept that group selection has played a significant role in maintaining sexual reproduction, it cannot have contributed importantly to the origins of sex; nor can it be important in determining the crossover frequency and recombination rates."

SICKLE-CELL ANEMIA

Hemoglobin S is an abnormal variant of hemoglobin A: persons heterozygous for hemoglobin S, having the constitution AS, suffer from a mildly disabling condition known as sickle-cell trait, in which the red cells give the appearance of collapsing into a crescent shape or of adopting other abnormal shapes under conditions of deoxygenation. Approximately a quarter of the progeny of parents both afflicted by sickle-cell trait will be homozygotes with the makeup SS. These are victims of sickle-cell anemia, a seriously disabling disease accompanied by gross abnormalities of liver and kidneys.

We may well ask how it comes about that the gene coding for transformation of hemoglobin A into hemoglobin S has not been wiped out by natural selection. The answer is profoundly interesting. The gene is prevalent in parts of the world where malaria is or was endemic—especially West Africa and the Mediterranean basin. The gene persists because the heterozygous victim of sickle-cell trait enjoys a measure of resistance to one of the severest forms of malaria, that which is caused by *Plasmodium falciparum*. This selective advantage is sufficient to make good the loss of the gene through death of the homozygous victims of sickle-cell anemia.

We have here a classic example of polymorphism maintained by the selective advantage of a heterozygote. It provides also an admirable example of the efficacy of euphenics, for if the environment is so far improved as to remove the threat of malaria, the gene coding for hemoglobin S will lose its selective advantage and over many generations will slowly disappear. Indeed, this appears to be happening in the U.S. South. The notion that improvement of the environment will necessarily lead to genetic deterioration can in this instance be seen for what it is—a bogey.

SOCIOBIOLOGY

Sociobiology is a candidate science, of which Edward O. Wilson's well-known text *Sociobiology: The New Synthesis* (Cambridge, Massachusetts, 1975) may be said to mark the instauration. At its highest valuation sociobiology might be regarded as a realization of John Stuart Mill's ambition (perhaps foremost among those that prompted him to devise a system of logic) to lay the foundations of a science of society. For sociobiology is indeed an uncompromisingly deterministic interpretation of social behavior.

Wilson's *Sociobiology* is reductionist to a degree that would have pleased Mill, who in his *System of Logic* wrote: "The laws of the phenomena of society are and can be nothing but the laws of the actions and passions of human beings gathered together in the social state ... Human beings in society have no properties but those which are derived from, and may be resolved into, the laws of the nature of individual men."

It is implicit in sociobiology—or is in any event its principal long-term ambition—that human social behavior and social usages (including perhaps refinements of the social contract) can be shown to have evolved, necessarily, by natural selection. As Wilson has it, "The pervasive role of natural selection in shaping all classes of traits in organisms can fairly be called the central dogma of evolutionary biology." Altruism is seen as the "central theoretical problem of sociobiology," for the very good reason that it poses in its most acute form the perplexing question of how it can be that genes or organisms can secure a net reproductive advantage through any form of behavior which, whatever advantage it may confer upon others, disadvantages itself. For some types of altruistic behavior—parental care, for example—there can be no doubt that kin selection offers a possible solution.

It is the lot of pioneers—and E. O. Wilson is no exception—that no matter how great the credit they receive for their theoretical innovations, they get blamed for the miscarriages and misapplications perpetrated by their followers. Enthusiastic amateur sociobiologists who do not fully understand the genetic theory of evolution by selection have propounded a whole array of notions—among them group selection—which do not stand up to critical examination. These turn upon the wish-fulfilling illusion that genes will become fixed in a population if it is desirable they should do so "for the benefit of the species" or the group. (See Richard Dawkins, *The Selfish Gene* [Oxford, 1976].) The trouble with the books propounding these views, according to one of their severest critics, is that "their authors have got it totally and utterly wrong."

This is not the only way in which the simplistic notions of geneticism make themselves apparent in sociobiology; other manifestations are the habitual tendency to depreciate the importance of exogenetic evolution and to attribute to the action of genes the determination of character differences that are more usually and perhaps more properly attributed to the influence of cultural heredity (in effect, doctrination and learning). Another manifestation, to be found also in sociolinguistics, is the belief that any character trait possessed by all the members of an interbreeding community—by all human beings, for example—must be genetically programmed.

From the standpoint of the history of ideas, sociobiology belongs to a lineage of thought that began with Herbert Spencer's majestic *System of Philosophy,* especially the *Principles of Sociology,* of which Benjamin Kidd (*Social Evolution,* New York, 1894, p. 2) wrote: "So little practical light has the author apparently succeeded in throwing on the nature of the social problems of our time, that his conclusions are held to uphold two diametrically opposed conceptions of social organisation, the individualist and the collectivist." Kidd did not believe that there exists a "science of human society properly so called" (p. 1); nor did Leslie Stephen (p. 5n).

It may be that sociobiology will prove itself in the interpretation of the social behavior of lower animals and will be looked back upon some day as a turning point in the history of social thought. However, it has not achieved this recognition yet; nor can it yet be said to be, as Wilson claims, a "branch of biology . . . to be ranked as coordinate with such disciplines as molecular biology and developmental biology."

SOMA

Meaning the body considered generally and as a whole, the word "soma" is most often used with an expressed or implied distinction from the reproductive cells that constitute the germ plasm and from the mind and spirit, the *psyche*. In educated conversation the word will most often be used with one or the other of these antitheses in mind, and the topic will frequently be the nature and the degree of the interaction between them.

Take soma and germ plasm first. August Weismann is chiefly remembered for his uncompromising insistence upon the separateness of soma and germ plasm—for his claim that no interaction can occur that would make possible the imprinting upon germ plasm of changes arising in the soma: acquired characters, whether passive or adaptive in origin, could not become coded in the genome. So readily is this argument accepted by nearly all professional biologists today that the character of the interaction between germ plasm and soma is no longer a matter of lively discussion.

A debate of somewhat different character—predominantly linguistic—has grown up in the past twenty years around the interaction of soma and psyche. In its original form this controversy grew out of the familiar behaviorist contention that there is not, or need not be, anything centered in our being to which the word "psyche" can usefully be applied. States of mind are known only by their overt behavioral manifestations, and these are what we should confine our attention to: it is almost always a work of supererogation to make reference to states or acts of mind. This is not exactly a view that has won universal acceptance; in fact, most thinkers have repudiated it with varying degrees of indignation or contempt. Modern linguistic philosophy is philological behaviorism: words are no more than their usages, which hold the key to all sense and meaning.

The questions about soma and psyche that constantly recur are, What is the relationship between mental states, thoughts, and other acts of mind and the brain considered as a physical object with its nervous interconnections and busy traffic of impulses? Is thought a secretion of the brain (as Darwin at one time thought it might be), or are thinking and the physical performances of the

242

SOMA

brain entirely different in character so that there can be no question of an interaction between the two? Gilbert Ryle's tremendously influential *Concept of Mind* (London, 1949) led to the widespread adoption of the view that propositions describing acts and states of mind belong to a different semantic category from those that describe the physical performances of the brain. It was therefore virtually meaningless to speak about a physical basis for acts of mind or about acts of mind affecting physical behavior.

We believe that this opinion was at least in part inspired by a wish to put down scientists and to quell any ambition they may have entertained to interpret mind in terms of matter and thus usurp some of the territory traditionally occupied by "philosophers of mind." Nevertheless, Ryle's argument is not very strong. Ryle describes it as a category mistake to suppose that a behavioral performance can represent the expression of an act of mind such as will or intention. He is using Edmund Husserl's notion of a "semantic category mistake" in a sense that departs from the original, for Ryle's category mistakes are simply mistakes, not semantic category mistakes (refer to P. B. Medawar, *Pluto's Republic* [Oxford, 1982], pp. 9–10).

Although the idea of psychosomatic medicine was at one time frowned upon by old-fashioned medical men, it is now regarded as a truism that states of mind can influence states of health and vice versa. There is even talk nowadays of a psychosomatic element in susceptibility to cancer—as there was at one time freely acknowledged to be in susceptibility to tuberculosis.

As to the interaction of mind and matter, consider a phenomenon such as *blushing*. No event could be more justly regarded as "mental" than the recollection of a past discomfiture, nor could any event be more justly regarded as physical or physiological than the reflex closure of the arteriovenous anastomoses that floods the capillaries of the skin with blood and causes blushing. Yet it is a truth of introspection that a person of refined sensibilities will blush on the recollection of some occasion of embarrassment such as a social solecism or a loud and confident mispronunciation of a word with which the speaker thought himself familiar. Here surely is an example of the effect of something purely mental upon a purely physical bodily activity.

No *serious* problem has grown out of the commonsensical ten-

243

dency to treat acts of mind like volition and conation as agents that bring about various behavioral performances. Medical scientists use the word "iatrogenic" to refer to disabilities that are the consequence of medical treatment. We believe that some such word might be coined to refer to philosophical difficulties for which philosophers are themselves responsible. To this category, whatever it may be called, belong most of the problems that have to do with the interaction of psyche and soma.

SPECIFICITY

If an estimate were to be made of the technical terms most frequently used in the professional writing and dialogue of biologists, we have little doubt that "specificity" would come out way ahead. Specificity refers to the complementarity or matched oppositeness found in reactions between antigen and antibody, enzyme and substrate, and between stimulus and response in reflex action.

SPINA BIFIDA

Spina bifida is a developmental abnormality of the embryonic axis in which neural folds fail to meet posteriorly across the dorsal midline to complete the closure of the neural tube. The result is that the communication between the cavity of the neural tube and the primitive embryonic gut remains open; a further consequence is that in this area the spinal rudiments fail to meet in the midline, leading to a bifid condition of the spine.

Anencephaly with spina bifida is so debilitating that many physicians question the wisdom of taking extreme—and in some countries extremely costly—measures to preserve the life of the newborn so afflicted. Thus spina bifida poses in a particularly acute form the deeply troublesome problems pertaining to unconditional preservation of life at birth. These problems take such widely different forms in the contexts in which they arise that it is extremely doubtful that any one pronouncement or act of legislation can cope justly with them all.

SPONTANEOUS GENERATION

Most natural laws are expressed in an affirmative form, that is, they are statements that certain natural performances or conjunctions of events under certain conditions *will* occur. Nearly all such laws can, however, be rephrased in such a way that, instead of being affirmative, they *prohibit* the occurrence of certain performances or conjunctions of events. So it is with the Law of Biogenesis: though usually expressed in the form epitomized by the familiar aphorism *Omne vivum ex vivo* (All that is alive came from something living), it can also be construed to prohibit the occurrence of spontaneous generation, which would be a flat repudiation of the law.

It is now the consensus that spontaneous generation does not occur—a belief so firmly held that any claim that it did occur (as was made by a Russian woman biologist in the 1950s) would be met with total disbelief.

This categorical disbelief is the product of very slow growth. Without knowledge of the doctrine of genetic continuity, the idea that maggots, mice, and worms could be generated by nonliving matter was not ridiculously farfetched. What happened in the course of time was that claims for the abiogenetic origin of relatively large and complex animals like frogs and mice were quietly abandoned, until in the end bacteria, fungi, and microorganisms (including protozoa) were the only creatures that it was thought might arise spontaneously. At one time it was believed that organic compounds, construed in a wide sense as compounds found in living organisms, could be formed only within such organisms. The natural or artificial synthesis of organic from inorganic molecules is evidence that the Law of Biogenesis does not apply at the molecular level.

By the time of Louis Pasteur (1822–1895), the idea of spontaneous generation was restricted to the possibility that bacteria might arise de novo in body fluids and in nourishing broths. The importance of Pasteur's famous experiments was to show that many examples of the ostensibly spontaneous generation of bacteria were all open to an alternative interpretation—namely the contamination of nutrient fluid in flasks by microorganisms present in the air. It is best to put the matter this way, rather than to say Pasteur's experiment refuted the whole notion of spontaneous generation, because *absolute* disproof is as elusive as apodictic proof. In just the same way, Lamarckism is no longer believed in, for all the examples of ostensibly Lamarckian inheritance hitherto described are open to another more conventional interpretation.

For an authoritative account of the debate about spontaneous generation, consult John Farley, *The Spontaneous Generation Controversy from Descartes to Oparin* (Baltimore, 1977).

SPORES

A reproductive cell that is in effect a gamete of asexual reproduction, a spore differs from a regular gamete in that it possesses the usual or diploid number of chromosomes. Thus a spore may give rise to an adult by a sequence of ordinary cell divisions. The term "spore" is sometimes used to refer to an asexual reproductive cell with a very low water content, which is highly resistant to environmental circumstances that would be lethal to most living organisms and certainly to regular gametes. (These, because of their requirement for permeability by sperm, cannot have the highly resistant and tough coating of the spore.) Bacterial spores in particular may be so highly resistant that they are virtually unharmed by anything except prolonged exposure to superheated steam (which, being at greater than atmospheric pressure, may have a temperature higher than 100 ° C).

SWEATING

Sweat is a weakly salty fluid containing urea and other excretory substances. Produced by long, coiled tubules formed in development from the outermost layer of the skin, the epidermis, it is distributed over the general body surface as well as over exposed areas such as the face. Though sweat contains urea, its excretory function is not paramount; its importance is as a thermoregulatory mechanism. That the evaporation of water causes cooling is a commonplace domestic observation, and all organisms except those with few sweat glands (such as dogs and rabbits) use sweating—in effect, the latent heat of evaporation—as a means of lowering the temperature. Sweat, it should be noted, does not just dribble out of the openings of the sweat glands: it is actively expelled by muscle cells also derived from the epidermis.

It is a matter of some psychological interest that dermatologists distinguish between two kinds of sweating: thermal sweating— that which is stimulated by elevation of the body temperature and which serves to lower it—and "emotional" sweating—that which is caused by anxiety, fear, and embarrassment. Whereas thermal sweating is the work of ordinary secretory cells in the simple so-called eccrine gland, the kind of sweating that has been described as mental or emotional in origin is transacted by glands of a very different nature and distribution: these are apocrine glands, in which the act of secretion involves the discharge of, more or less, the substance of the cell. Apocrine sweat glands are formed over a wider area of the body than that in which they persist, and they tend to disappear from all except the so-called apocrine areas (specifically the armpit, mons veneris, anus, and mammary regions). Because it contains cell granules and fatty acids, apocrine sweat is responsible for the body's characteristic smell. It is a well-known theorem of boys' literature that dogs can smell fear, and it is a matter of common experience that human beings can smell embarrassment in others—a giveaway commonly annulled nowadays by use of an antiperspirant.

SYMBIOSIS

A mutually advantageous association between two unrelated organisms is known as symbiosis. The most familiar example is the symbiosis between fungus and green alga that forms a lichen: the composite organism can live wherever a fungus can live, and its versatility is enormously enhanced by being able to use atmospheric carbon dioxide as a plant does, so that lichens flourish on dry stone walls. Another example of symbiosis is the relationship between ruminant animals and the huge population of microorganisms that lives in and on the contents of the rumen, where they help to digest the otherwise indigestible food of their hosts (hay and other such delicacies) by breaking down cellulose into smaller, assimilable fragments.

The partners in a symbiotic relationship are referred to as

symbionts. Symbionts need not necessarily belong to different groups of organisms; but in the most spectacular cases of symbiosis they invariably do, for it is only when their natural capabilities are very different that the symbiotic relationship can be seen to work to special advantage.

It is easy to visualize how symbiosis might have evolved out of mere commensalism and might easily evolve into parasitism. Commensalism (literally, sharing the same table) is not necessarily mutually advantageous. More often it is an opportunistic association in which one animal's feeding habits and table manners are such as to provide a suitable environment for the other. Consider, for instance, members of the perch family that attach themselves by suckers to sharks or swordfish and live on the scraps that fall from the rich fish's table, the involuntary host getting no benefit and being unable to enjoy the pleasure a rich man may derive from the sycophancy of his beneficiary. Commensalism is especially well developed among species of annelids (true worms) which have developed a number of stratagems for living in the tubes or other shelters of unrelated organisms.

Parasitism too may be derived by the evolutionary extrapolation of symbiosis. In parasitism the advantage of the symbiotic relationship is no longer two-sided. The parasite derives all the benefit; the host derives none. But since it is clearly not in the parasite's interest to kill, the relationship does not as a matter of course lead to the death of the host.

SYMPATHETIC NERVOUS SYSTEM

As its alternative designation, autonomic nervous system, connotes, the "sympathetic" is that compartment of the nervous system which provides for the automatic, involuntary activities of the body. Until philosophers made us feel uneasy, we used to say activities such as beating of the heart, sweating, peristaltic movement of the gut, and all that is entailed by the notion of vasomotor control occur without the interposition of mind. The autonomic nervous system—under the guidance of its "brain," the hypothalamus—innervates the heart, the smooth muscle of blood vessels and intestines, and the sweat glands. The sympathetic and parasympathetic divisions of the autonomic nervous system are mutually antagonistic in their effect.

It is an anatomically distinctive characteristic of the sympathetic nervous system that its constituent neurones (the cell bodies of which are in the central nervous system) do not directly innervate their respective end organs, but do so through a relay: the nerve fiber issuing from the central nervous system makes synapse with a second neurone, of which the cell body usually lies in a sympathetic ganglion beside the spinal cord, though sometimes it may lie nearer the end organ. The former fiber is described as preganglionic, and the latter, which proceeds directly to the end organ, is postganglionic.

The synapse is responsible for appreciable delay in the propagation of a nerve impulse. It is that which polarizes the impulse, that is, provides that the nervous signal travel only one way (though it will travel both ways from a point of stimulation within a nerve fiber itself).

The discovery that the nervous system is cellular in structure may be accounted the greatest triumph of the cell theory. It was something of a conceptual triumph, too, because the cellular structure of the nervous system is far from obvious. Its recognition was not an ideological thunderclap but grew out of lengthy and scrupulous investigations by the Spanish neuroanatomist Santiago Ramon y Cahal (1852–1934) and the physiologist and histologist Edward Sharpey-Shafer (1850–1935).

A neurone resembles an ordinary cell in having a nucleus surrounded by a domain of cell sap, but differs from an ordinary cell in that the cell body is drawn out into very long extensions, the nerve fibers (axons, dendrites), which in large animals may be several yards long. The cell body that surrounds the nucleus is sometimes referred to as the perikaryon and sometimes simply as the cell body.

The relatively enormous length of neurones poses the difficult biophysical problem of how the cell nucleus controls the cytoplasm. The rate of diffusion is such that when a solute of constant external concentration is diffusing into a substance permeable by it, the distance traversed by the wave front of the diffusing substance is proportional to the square root of time. Control of synthesis at the extreme end of an axon cannot therefore be mediated through ordinary diffusion from the nucleus, and the idea has accordingly gained ground that there is a movement of cell substance down the axon—or conceivably that the axon is continuously growing out of the perikaryon.

The complexity of neurones is such that they are incapable of cell division; the loss of a neurone is therefore irreparable. It can, however, regenerate, the rule being that when an axon is severed, the peripheral part of it (that which lies away from the cell body) disintegrates and regeneration starts from the center. It is a slow business, and functional recovery after the severance of a peripheral nerve is as a rule rather poor: the right connections have to be made, and the inflammation that accompanies severance must subside; in addition, the newly grown fiber must be invested anew by an insulating fatty sheath of myelin.

The function of the long nerve fibers is of course to conduct nerve impulses. The tendency is to use the word "axon" to designate a nerve fiber that conducts in a direction away from the cell body and "dendron" for the perhaps shorter and stumpier fiber that conducts in a direction toward the cell body.

SYPHILIS

The Great Pox, a highly contagious disease caused by a bacterium of the kind known as a *spirochete,* is full of interest for the philosophically minded biologist. It is widely believed that syphilis was introduced into the Old World by the sailors returning from Columbus' great voyage to the West Indies, although of course one cannot be certain that the disease had never occurred before in Europe. A pox, after all, is only an eruptive disease associated with pustules and vesicles, and any case of syphilis that might have arisen could easily have been dismissed as one of the more familiar eruptive diseases. In due course it came to be understood that syphilis may be congenital, because it can be transmitted to the infant in utero (doubtless one of the factors contributing to the punishment theory of illness).

Syphilis has an up-and-down clinical history and may be covert for rather long periods. With all such diseases it is extremely difficult to evaluate the efficacy of various forms of treatment.

Paul Ehrlich (1854–1915), the brilliant founder of chemotherapy, introduced the compound arsphenamine under the trade name salvarsan and subsequently improved it. In spite of skeptical attacks drawing attention to the extreme toxicity of arsenicals, neosalvarsan was eventually accepted as a remedy. Of course, it was never subjected to the intensive controlled clinical trials to which therapeutic innovations today are exposed. The causative organism of syphilis is susceptible to a number of antibiotics including penicillin, and these have now superseded arsenicals in the treatment.

Spirochetes are so designated because their bacterial cell substance is wound helically around their central core or filament; they progress by a mixture of bending and corkscrewlike movements. Somewhat unlike the common run of bacteria, spirochetes were at one time classified as protozoa. In reality they contain a number of distinctively bacterial constituents, among them muramic acid.

The causative organism of human syphilis is the spirochete *Treponema pallidum.* In untreated treponematosis the organism is widely disseminated throughout the body, including the nervous system; it accordingly gives rise to a variety of sensory abnormalities (Beethoven's deafness is reputed to have been syphilitic in origin) and to symptoms mimicking various neurological diseases. The extreme end result is general paralysis of the insane, in which progressive dementia may be accompanied by tremors, progressive spastic paralysis of the limbs, and psychological disturbances such as delusions of grandeur. These awful symptoms, especially when thought to be retributive in character, deeply impressed and shocked nineteenth-century and fin-de-siècle writers.

TAXON

This is a comprehensive term for a classificatory group or taxonomic category, irrespective of rank, though with the extra connotation of validity in the sense defined below. An example of a classificatory term that is *not* a taxon is "invertebrata," the invertebrates: the term must be thought of as a description or comment, not as a taxon. The same applies to Aristotle's long-obsolete term "sanguinea" for the hodgepodge of animals related only by the possession of red blood.

The system of biological classification forms a hierarchy, and the fact that it does so is strong evidence in favor of the theory of evolution. The grandest genuine taxon in this hierarchy is Linnaeus' "kingdom" (*regnum*), as in animal kingdom or plant kingdom; lesser taxons are described and commented on below.

A *phylum* comprises organisms with a common ancestor having the same morphological ground plan as do chordates and arthropods (insects, crustaceans, millipedes, centipedes, scorpions, spiders, and king crabs).

Next in the hierarchy comes the taxon *class*. Classes are the bigger branches of the great tree represented by a phylum. The Vertebrata, for example (a subphylum), is composed of the following classes: fishes (pisces), amphibians, reptiles, birds, and mammals. The several classes of animals differ a good deal from one another in abundance and variety, but so far as judgment can make them so, the *order* that forms the next taxon is of the same stature throughout the animal kingdom. It is not possible to define precisely what degree of affinity justifies the bringing together of animals into an order: it is obviously higher than that which unites the members of a single class. This really is a matter of "feel," and the following examples illustrate the level of affinity aimed at: whales and dolphins on the one hand form an order Cetacea, and on the other hand seals, sea lions, and walruses form an order Pinnipedia. These of course are all mammals, but birds illustrate the same principle—though there is no strictly objective evidence that herons and storks (order Ciconiiformes) have the same degree of likeness to each other as walruses and seals. Even to a layman, though, it makes sense that ducks, geese, and swans should form one order—the Anseriformes—and that vultures, hawks, falcons, and ospreys should form another—the Falconiformes.

The next three ranks in the hierarchy are genera, species, and individual organisms. A *genus* comprises animals "of a kind" in the sense in which lion, tiger, panther, leopard, and jaguar are of a kind; considered severally, each is allocated to a different species, that is, they are regarded as being of a different specific kind. Thus in Linnaeus' two-termed nomenclature, and in the terminology of the Zoological Society of London, the four great cats just mentioned, members of the genus *Panthera,* are allocated to the respective species *Panthera leo, Panthera tigris, Panthera panthera, Panthera pardus*, and *Panthera jaguaris.*

At some stage of his life everyone who describes himself as a biologist has been called upon to worry over "the species problem"—that is, the problem of defining the taxon in terms applicable to all species. The problem is at least in part logicophilosophical and one that arises out of the need to reconcile the notion of an individual's membership in a species with the geneticist's inclination to define a species in terms of a characteristic configuration of genes. It is a real problem: a taxonomist wants to be able to say of any one organism that it is a member of such and such a species—a judgment that will not come readily to his lips if he has already been driven nearly out of his mind by assurances that a species is essentially a cluster of points in n-dimensional character-space.

The dilemma can be illustrated by the example of a Pennsylvania candy maker who puts up chocolates in several distinctive packs: the Philadelphia pack has the following flavors in the percentages stated: mint (50 percent), coffee (20 percent), orange (20 percent), pear (= amyl acetate, 10 percent); whereas the Pittsburgh pack has the same allelic flavors in the percentages 40, 30, 20, and 10. Given this statistical flavor-frequency definition of the several candy packs, we could not say positively of any specific piece of candy that it clearly came from one pack or the other, yet this is what the museum taxonomist is called upon to do in passing judgment on species membership.

If species are defined as some population geneticists would like, in terms of gene frequencies, then only *populations* of animals can be allotted to species. The population-genetic definition of a species is not in fact so very far removed from the commonsensical. A distinctive pattern of gene frequencies implies that by some means or other—genetic, geographic, or behavioral—the species enjoys a measure of reproductive isolation, without which the distinctive pattern of the species gene pool would soon be lost.

The purpose of a classification is to provide nomenclature and give order. All classifications are of course founded upon the assessment of degrees of affinity; if a wholly arbitrary scheme of classification were indeed possible—if, for example, an arbitrary taxon made up of "animals with faces" or "animals that fly" were invariably to gather into one group animals also related in *other* ways—then we should have real reason to question the hypothesis of evolutionary relationships. But no such wholly arbitrary judgment can be undertaken: if we do classify animals by their affinity, the resulting scheme falls into that hierarchical arrangement which is only intelligible on the premise of descent with modification. Consider for example the classification of a familiar occupant of the nursery.

Phylum:	Chordata
Subphylum:	Vertebrata
Class:	Aves
Order:	Anseriformes
Genus:	*Anser*
Species:	*Anser disneyi*
Member:	Donald Duck

For a taxon to be valid, a common ancestor of the members of the taxon (or its nearest modern representative) must itself be a member of the taxon. Otherwise the taxon will be judged polyphyletic (comprehending two or more lines of descent). The class Reptilia is a case in point: the animals living or extinct that are classified as reptiles belong to two major lines of descent, one leading to mammals and the other to birds—a dichotomy that began as long ago as in the amphibia.

There is in principle no reason why there should not exist a taxon lower in the hierarchy than species—that is to say, a population of animals with a distinctive genetic makeup that forms a subclass of the species as a whole and that enjoys a degree of reproductive isolation which preserves racial characteristics. When the term is used of populations reproductively isolated by impassable barriers, then we may properly speak of "races"—as in island races of mice or other small animals. If usage were confined to this context it would cause no offense, but the word has unfortunately offered an irresistible attraction to illiberal mischief makers intent upon making a political or a cultural point, especially one based upon the presumed or postulated inborn inferiority or superiority of some one race. This usage must be judged inadmissible. No free human community combines a distinctive genetic makeup with a degree of reproductive isolation sufficient to maintain it.

If all Jews, for example, were strictly assortative in their choice of mates and never married outside their religion, then it could be said that some degree of reproductive isolation had been achieved by cultural means. Such a state of affairs is so far short of fulfillment that use of the term "race" with reference to Jews is as ill judged and unsound as the adoption of such an expression as "the island race" with reference to the British. For Britain has been the subject of repeated invasions and immigrations since the original invasion by the English in the fifth and sixth centuries; its people, politically speaking, are as hybrid and as mixed in origin as any other. The English are no more pure than their language, which is quite wonderful in the multiplicity and diversity of its roots.

Our recommendation, accordingly, is that the term "race" be abandoned except in those technical biological texts where there is a prima facie case for its use and where its taxonomy can be critically appraised.

TELEOLOGY

"Four causes underlie everything: first the final cause, that for the sake of which a thing exists." So wrote Aristotle in *De generatione animalium*. Teleology is the "science" of final causes and has to do with their ascertainment and interpretation. Final causes have a special importance in biology because, being adaptive in character, evolution has so often worked as if teleology exercised, as Aristotle thought it could, a sort of causal traction.

We believe that with the important qualification noted below there is nothing intrinsically offensive in the notion of teleology. It is folly or ignorance to deny that the purpose of nests is to protect the relatively helpless young of birds and mammals, and of the amnion to provide the embryos of land vertebrates with the aquatic environment they need in order to develop. The purpose of teeth, moreover, it can now be revealed, is mastication; of eyes to see, and of ears to hear.

What is wrong, then, with teleology and why do biologists repudiate it with such fastidious distaste? What is wrong with teleology, as Aristotle explicitly conceived it, is the use of purposes as causal explanations of phenomena, as if the need for some biological structure was reason enough for its coming into being.

Teleology has intruded to some extent into evolution theory, and into sociobiology in the form of the aberration of thought John Maynard Smith describes as Panglossism. A typical example would be that evolutionary changes can come about "for the good of the species," and that this is explanation enough for their occurring.

To avoid these unacceptable connotations, the genteelism "teleonomy" has been coined. Teleonomy is teleology purged of all its pretensions to providing causal explanations, and restricted exclusively to putting on record the purposes which biological structures and performances do in fact fulfill. The word "teleonomy" has not caught on, perhaps because corruption of biology by teleology is not now so grave or so imminent a danger as it was once feared to be.

TERATOLOGY

Embryonic development is a highly complicated process and many of its episodes—especially the formation of the embryonic axis—are extremely vulnerable to disturbances. The process of development itself may compound and amplify the effects of these disturbances: thus failure of the neurenteric canal to close may lead to a bifid condition of the spine, and this in turn to other grave abnormalities. These embryologic accidents and misadventures may lead to the birth of physically handicapped and structurally deformed children, "monsters," who are much more numerous than is commonly supposed.

The classification and investigation of these abnormal embryos and newborns are the subject matter of "teratology," a word that suggests pretensions to the stature of a science (a designation not really deserved). The formation of monsters such as Siamese twins is an interesting common territory shared by pathology and embryology, but teratology has not—as had at one time been hoped—thrown a flood of light upon developmental processes, and it has not helped us very notably in the interpretation of normal development. Teratology is more deeply in debt to embryology than the other way around.

TERMINOLOGY

Although some of the terminology of biology is of the functional, workaday kind we expect of a complex and busy science, all too much of it is a mixture of grandiloquence and linguistic offenses. We shall start with the philological chamber of horrors known as immunology. "Antibody" would not win a prize for horribleness; it is a perfectly acceptable formation for a living language. But "antigen" has nothing going for it; medical students are known to feel rather embarrassed when they first hear it—a reaction that reveals in them sensibilities finer than they are reputed to possess. But we shall not attack a terminology that no one defends. "Tolerance" as a description of a specifically nonreactive state is poor because it is not a word that lends itself to inflection, and the words "tolerize" and "tolerogen" have only the sentimental appeal of keeping alive a long-standing tradition of terminological ineptitude in immunology. Still, peace of mind can be bought with not very lengthy periphrases: "to make tolerant" or "tolerance-conferring agent" suggest themselves. Such niceties would be dismissed as bourgeois pedantry by those left-wing hotheads who believe that this kind of fastidiousness is inimical to the welfare of the working classes.

The floridity of much of the terminology devised by biologists in the nineteenth century may have the same origin, we believe, as the tortuous quasi-scientific locutions of modern sociologists: it was designed above all to impress. Biologists of that era were not very sure of their ground and probably thought they were not receiving the esteem from the general public and from fellow scientists that they felt was their due. In the nineteenth century physicists and the distinguished civil engineers who built tunnels, bridges, engines of war, and the first great ships were the people most revered; biologists, as scientists, were sometimes treated with a kind of condescension against which they sought to defend themselves.

Two favorite pretensions may be cited: the mobile protrusions of the body by which simple creatures such as amoeba creep around the substratum could not be mistaken for feet by anyone in full possession of his faculties; it was hardly necessary, therefore, to designate them as *pseudopodia* ("false feet," the elementary texts explain). An international commission such as that which ruled on the nomenclature of human body parts might now rule that these tiny organs should henceforward be called *podia*. The same considerations apply to the designation of a reproductive element in the life history of a parasite of earthworms. With the kind of license allowed those ancients who discerned the outlines of swan, bear, dog, or warrior in the pattern of the stars, these reproductive elements were designated *pseudonavicellae*. We can only hope that a huge accession of self-esteem followed the nomenclatural folly of describing these small reproductive elements as "as-it-were-little-boats."

Sociologists today are looked on much as biologists were in the nineteenth century, and this may help to excuse their egregious terminology.

THERMOREGULATION

The maintenance of the temperature of the body within functionally suitable limits is known as thermoregulation. In many animals it is achieved behaviorally, by avoiding situations that will make the individual too hot or too cold. This is of course not possible for animals that enjoy a wide range of habitat and freedom of movement. For them—and this applies especially to land vertebrates—the problems vary from animal to animal, depending on size and shape, and they are surmounted in many different ways. The problem for very small warm-blooded animals is conservation of heat; because of their size they have a relatively high surface area in relation to their volume (pigmy shrews, for example, may weigh only 5 grams as adults and have an acute problem in maintaining body temperature, one that is surmounted by virtually continuous eating). In such tiny animals fasting for half a day can lead to death from starvation. They are still more vulnerable when young, and this difficulty is surmounted principally by nursing and by the thermally well-insulated environment of the nest.

More generally, though, the real problem of thermoregulation is excretion of heat, and this particularly afflicts large mammals with dark skins. Such mammals often have a thick layer of blubbery fat in the skin, which compounds the difficulty. Being dark, moreover, they tend to absorb all the infrared radiation that is around (every motorist has observed that a dark car gets warmer in the sun than a light-colored car). For such animals lowering the temperature is partly behavioral: hippos, we know, live partially submerged. It is difficult to see how they could survive otherwise, for although we have been conned by etymology into thinking of the hippopotamus as a river horse, it is in fact a river pig, and like others of its kind is extremely fat. The African elephant can increase its body area by a sixth by spreading its huge ears

Furry animals like dogs and rabbits that are poorly equipped with sweat glands do nevertheless manage to make use of the cooling effect of the evaporation of water: there is substantial heat loss from the lungs, and this a dog supplements by evaporation from the tongue in its characteristic panting. Rabbits have relatively large and highly vascular ears which help in thermoregulation. Human beings lower their temperature by sweating and can control their temperature by various exosomatic devices like clothes and shelters.

TISSUE CULTURE AND STORAGE

The procedure whereby living cells are preserved outside the body in a germ-free nutrient medium is called tissue culture. If the culture media are suitably chosen, the cells can continue their normal physiological performances: heart muscle will keep on contracting rhythmically, cells capable of division will divide, and nerve cells will continue to sprout those slender extensions along which, in vivo, nerve impulses are propagated.

The tissue culture technique was introduced by Ross Granville Harrison of Yale University (1870–1959), one of the greatest American biologists. He cultured fragments of the central nervous system of frogs in drops of lymph and was able to watch the outgrowth of nerve fibers. For many years afterward "hanging-drop cultures" were the routine means of cultivation; a drop of nutrient fluid was spread on a cover glass and inverted over a hollow ground slide, so as to make microscopic inspection from the top easily possible. Surface tension was usually enough to hold the drop in place, an end facilitated by using a clotting fluid such as blood plasma or lymph in the nutrient medium. When cells capable of division were used (connective tissue cells from the heart of a chicken embryo were a favorite choice), the cells continued to divide; when the tissue fragment had enlarged sufficiently, it could be divided into two and each half subcultured.

Alexis Carrel and his colleague Albert Ebeling of the Rockefeller Institute for Medical Research worked with tissue culture on a large scale, using what can be seen with hindsight to have been an unnecessarily elaborate ritual. This created the widespread impression that tissue culture was altogether too esoteric and complicated for the routine use of experimental biologists working in ordinary down-at-the-heel laboratories; but, as usually happens, familiarity with the procedures tempted experimentalists to be less cautious. One risk the pioneers had been at pains not to run was allowing the fragments of tissue for culture to drop to room temperature before they were explanted into their culture media at body heat. Experience soon showed this precaution to be unnecessary. Tissue fragments remain alive even at room temperature if submerged in the kind of fluid that will support their growth in tissue culture. Indeed, the storage of tissue fragments for a day or two in a refrigerator before culturing became a quite ordinary practice. Alternatively, cultures could be prepared for incubation and then kept cool (5–10 ° C) until it became convenient to put them into an incubator at body heat, whereupon growth would start.

The culture medium most frequently used was a mixture of lymph or blood plasma and embryo tissue juice. The lymph or plasma formed a firm fibrin clot especially suitable for the outward migration of cells, and the embryo juice contained all the nutriments required to maintain cellular movement and division for several days.

It was at first believed that tissue cultures were "immortal"—or to put it more exactly, that the division lineages to which such cells gave rise were indeterminate. More careful studies have shown that a culture of *normal* cells (cells that have not for any reason undergone a malignant transformation) has a determinate life span and dies out after a finite number of divisions, the number depending upon the chronological age in vivo of the cultured tissue when the culture began. The famous and long-lived Rockefeller strain of chick heart fibroblasts therefore must have been refreshed by the accidental recruitment of new cells during the cultivation process, perhaps from the embryo juice used to nourish the cultures. Or, of course, the cultures may from time to time have been started afresh by acolytes unwilling to incur the odium of being thought to have lost a strain that was rapidly achieving the status of an heirloom.

The early days of tissue culture were in some ways unsatisfactory: culturists were so beguiled by the beauty of cultured cells and by the miracle of its being possible to use them at all that they did not utilize them for study. The result was a tendency to dismiss tissue culture for serious biological problems as a hobby. In spite of these criticisms, much of high interest was learned about the behavior of cells, especially about the mechanisms of cell movement and division and about phenomena such as pinocytosis, the imbibition by cells of tiny droplets of fluid from the external medium. Tissue cultures, moreover, have some quite instructive organismlike qualities. It was observed, for example, that when a tissue culture had grown to its maximum size as a circular disk of cells, the excision of a sector in the shape of a cake slice was followed by regeneration until the gap filled in and the circular perimeter of the culture was restored, a phenomenon not explained until the process of contact inhibition was recognized. All these advances were made possible by the ease with which hanging-drop cultures lent themselves to high-powered microscopic investigation.

There was a tremendous revolution in culture methods when cultured cells were grown by essentially bacteriologic techniques: in the United States and Canada cells began to be grown in fluid cultures, the individual cells not necessarily attached to a substratum such as the glass or plastic surface of the culture vessel. This was a particularly suitable method to cultivate viruses for the preparation of vaccines, or to harvest cellular products such as interferon. And so tissue culture entered an era of semiscale technology: there was indeed even talk of steaks grown in vitro. A moment's reflection on the input of nutrient in relation to the output of cells shows the ambition to be wholly unrealistic—or rather, one much better achieved by such homely methods as feeding cattle with corn and thereby making it possible for them to convert plant protein into animal protein.

With the growing success of the transplantation of organs and tissue, storage advanced to the forefront of technical problems in applied biology. Tissues vary greatly in toughness as measured by the ease of storage: not unexpectedly the outermost layer of the skin, the epidermis, is the toughest and most resistant tissue in the body. Skin can be stored for a week in the refrigerator, with precautions no more elaborate than avoidance of drying. Storage for periods of months requires attention to detail. It was discovered by accident at the National Institute for Medical Research in London that glycerol is remarkably well able to protect tissues against the damage that would otherwise be caused by freezing them solid. Thus when thin skin slices like those used for grafting are impregnated with a 15 percent by volume mixture of glycerol with a balanced salt solution, they can be stored in a sealed vessel for months at −80 ° C, or for a period of years in liquid nitrogen. From such observations as these grew a cadet branch of biology, named cryobiology from the Greek root *kryos,* icy cold.

Unfortunately, the storage of organs such as hearts and kidneys is a matter of far greater difficulty and raises problems that have not yet been solved (though we need not doubt that they will be). Reports of the deep-freezing of mammals the size of rats and hamsters have led to the too-easy assumption that the techniques of keeping alive organisms as large as human beings have already been mastered. They have not. In our opinion, money invested to preserve human life in the deep freeze is money wasted, the sums involved being large enough to fulfill a punitive function as a self-imposed fine for gullibility and vanity.

After storage by freezing the next innovation to be expected is storage by drying, an ambition fueled by the rather special example of the crustacean *Artemia salina,* the eggs of which can be stored in a vacuum desiccator in the presence of phosphorus pentoxide—that is, in complete dryness. Matters have not gone further, however, than the choice of the word *xerobiology* as a suitable name for this branch of human endeavor.

TOXINS

The word "toxin" usually refers to a poisonous substance of biological origin which as a rule is toxic by reason of its physiological action. Toxins are more poisonous by several orders of magnitude than conventional chemical poisons (toxicity is measured by the dosage in relation to body weight that is necessary to kill or intoxicate one-half—"the median lethal dose"—of all organisms exposed to it). Microorganisms that produce substances toxic to other microorganisms (namely antibiotics) form a class apart; the most notorious producers of substances toxic to animals and to man are bacteria, fungi, snakes, and some fish.

The toxins produced by bacteria are of two kinds: exotoxins are liberated into the environment from the bacterial cell, and these are especially virulent when the bacteria that generate them invade the tissues. Such are the toxins produced by the organisms giving rise to diphtheria, dysentery, gas gangrene, and botulism. Botulism is normally contracted by eating foodstuffs in which the organism has grown, because its instigator, *Clostridium botulinum,* does not multiply in the tissues. When toxins are an integral part of bacterial structure, they are referred to as endotoxins, which are liberated only upon disintegration of the cell. These endotoxins, composed as they are of protein, polysaccharide, and fatty matter, are probably the most complicated chemical substances known.

The toxins manufactured by bacteria, mushrooms, fish, and snakes do not act in any one way but in a whole variety of different ways. Many are enzymic, and some act by promoting or preventing the clotting of blood. Snake venoms of this sort have some clinical uses in the treatment of hemophilia. Many toxins cause acute abdominal pain, accompanied by nausea and vomiting. Others, greatly prized for this property, arouse hallucinations; many are generally neurotoxic. Hemlock, for example, causes muscular weakness leading to paralysis of the limbs and eventually to respiratory paralysis and death. The symptoms of strychnine poisoning are familiar to readers of detective novels, who have been shocked to learn that it produces tonic convulsions (muscular contractions long sustained instead of "clonic" convulsions regularly repeated), muscular rigidity, respiratory distress, paralysis of the heart, and death.

Animals of all kinds can produce toxins. The toxin of the Japanese puffer fish is said to be the most poisonous substance known, a distinction previously held by the toxin of *C. botulinum*. The fish is nevertheless eaten in Japan after preparation by experienced chefs. The first symptoms produced by the toxin are tingling, then a paralysis of lips and tongue accompanied by difficulty in swallowing. It is said that to experience symptoms of this degree is an amusement of young bloods in Japan. There is no known antidote, so having gone thus far it is advisable to abstain; no valid philosophical point is made by allowing the symptoms to proceed to general numbness, muscular paralysis, and death.

TRANSFORMATIONS

As explained in the entry on FORM AND MATHEMATICS, the method of transformations makes possible the mathematical analysis of change of form as opposed to the mathematical representation of form itself. The method was first used in a zoological context by Sir D'Arcy Wentworth Thompson (1860–1948), the natural historian and classicist who held the chair of natural history at Dundee University and later at St. Andrews, for sixty-five years. He endeavored to compare outline shapes of organisms, not bit by bit or character by character but as a *whole*. A single example illustrated below makes the character of the method intuitively obvious: it is the transformation that assimilates the outlines of the fishes *Diodon* (left) and *Orthagoriscus* (*Mola*) (right).

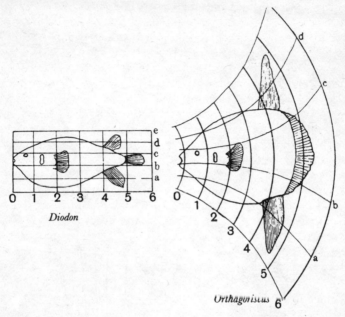

Several other examples are to be found in Thompson's famous essay *On Growth and Form* (Cambridge, 1917): two are shown here.

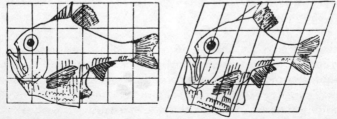

The mathematics of Thompson's little morphological adventure has several points of interest, but before propounding them

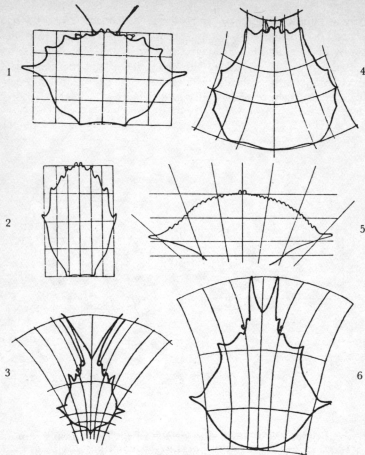

Carapaces of various crabs. 1, *Geryon;* 2, *Corystes;* 3, *Scyramathia;* 4, *Paralomis;* 5, *Lupa;* 6, *Chorinus*

we should clarify a few technicalities. All the principles of the method of transformations can be explained with reference to a plane or two-dimensional system, and to this we shall confine ourselves. Any plane figure can be represented as a set of points, each point being a number couple representing a pair of corresponding values of two coordinates x and $y;$ membership in the set is specified by an algebraic formula or function from which corresponding values of x and y can be read.

We might, for example, devise a function

$$y = \text{circle } (x). \tag{1}$$

This expression can be taken to mean that x and y are so related that their values will define the outline of a circle when plotted on conventional graph paper, as shown on the left below. Suppose we now wish to transform this circle into the larger circle shown on the right. There are two ways to do this: by a change of points or by a change of coordinates (in practice, both amount to the same thing).

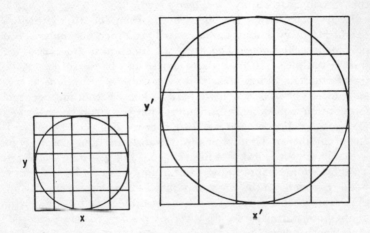

To represent the transformation as a change of coordinates—as a change of space rather than as a change of points—we must in the D'Arcy Thompson manner devise a different space frame. If the new coordinates are twice their original length, the two space frames are as shown in the diagrams. The relationships between the old coordinates $[x, y]$ and the new ones $[x', y']$ are given by the equations

$$x' = 2x, \qquad y' = 2y. \tag{2}$$

The consequence of this is that if we plot our old circle formula (1) on the new grid, we get a circle with four times the area of the original one. Alternatively, we may represent the transformation as a change of points and keep the old function (1) and the old grid and simply substitute the new points x' and y' so that formula (1) now becomes

$$\frac{y'}{2} = \text{circle}\left(\frac{x'}{2}\right). \tag{3}$$

If, now, equation (3) is plotted on the original grid (at the left), the result will be the same as plotting equation (1) on the transformed grid (right).

The equations (2) that are the key to this simple transformation and that map the old circle into the new one are known, from their use in cartography, as *mapping functions.*

Even when restricted to the two-dimensional outline drawings to which D'Arcy Thompson necessarily confined his analysis, it is possible to devise an explicit algebraic function such as (1) to define the outline of a fish or whatever else may be the subject of the transformation; but *the use of this method is in no way dependent upon the possibility of doing so.* All we need to know is that an algebraic formulation is possible in principle—that there can be conceived to exist a function such as $y = \text{fish}_1(x)$ or $y = \text{fish}_2(x)$, to cope with one or another of Thompson's fishy transformations—or, to cope with still another kind, $y = \text{man's face}_1(x)$ or $y = \text{man's face}_2(x)$. To these we may apply any transformation we please, the point (made especially in the entry on FORM AND MATHEMATICS) being that although it will not normally be possible to define shape with mathematical exactitude, it is certainly possible to define *change* of shape mathematically.

D'Arcy Thompson's method of transformations is open to a number of objections of some weight. In the first place comparison of adult with adult is rather unbiological, because the transformation is not one that could take place in real life. It is a pity Thompson did not have figures that would have enabled him to compare a juvenile with an adult form; but perhaps if he had done so, a second shortcoming in the method would have become increasingly apparent: the fact that his method is entirely static in character. D'Arcy Thompson essentially is comparing two stills from a cinematic film.

Despite all its faults—a consequence of the extreme difficulty of the project—the method of transformations does make a valid biological point: in differentiating forms, it is not always necessary to compare them point by point and detail by detail, calling attention to an enlargement here and a diminution there, a broadening in one place and a narrowing in another. It may be that we can envisage the transformation as the consequence of one comprehensive morphogenetic process, such as might be the work of a single gene. This important and illuminating conception is Thompson's real contribution to morphology.

Unless the method is to remain completely sterile, it must be put into motion—it must be made kinetic instead of static—and in particular it must be applied to development, or at least to a sequence of evolutionary changes. Change of shape must be represented as a function of age or secular time. The mapping functions are the key; and the principle of the method is to gear the transformations by parameters that vary in dependence on time, whether geological or developmental. If we stay with the simple example shown on page 267, we must first express the mapping functions (2) in a more general form:

$$x' = k_1(x), \qquad y' = k_2(y). \tag{4}$$

Given the mapping functions, it is clear that the values of the two parameters k_1 and k_2 determine the form the transformations take (that is, the exact degree of enlargement or—if the degree of magnification in the two dimensions is not the same—the degree of distortion to the shape of, for example, an ellipse).

The key to putting the transformation into motion and converting these lantern slides into a cinema film lies in the mapping functions (4): as we have said, the parameters must be deemed to be variable and to vary in dependence upon time. We may write therefore:

$$k_1 = f_1(t), \qquad k_2 = f_2(t). \tag{5}$$

If these functions represented relationships of simple proportionality, then clearly the rate in time of enlargement or of distortion would be constant, and in uncomplicated examples one might hope to be able to represent the form of an organism as a function of its age. The few, only moderately successful attempts to represent change of form as a continuous parameter transformation of this kind justify the inferences that change of shape is a monotonic function of age and that the rate of change of shape is higher earlier in life than later on, tending progressively to slow down. There is a wide open field for research here, but it would have to be based upon original measurements, photographs, or observations: no ready-made data can be readily adapted to analysis by continuous parameter transformation.

TRANSPLANTATION

The act of moving an organ or member or tissues or cells from one organism to another, transplantation is a useful procedure in horticulture and arboriculture. In human beings and other vertebrate animals it is complicated by surgical and sometimes physiological difficulties, and by the intrusion of immunologic factors having to do with incompatibility.

The transplantation of skin illustrates both the procedure and the terminology that goes with it. Although a superficial injury of skin heals of its own accord quickly and effectively, no regenerative process can make good a loss of the full thickness of the skin by burning, for example, or by some other accident. Natural healing involves contraction, the gathering together of the margins of the wound by a process that often causes disfigurement and functionally disabling distortions of surrounding tissue. The only definitive remedy is skin grafting, a procedure in which a thin slice of skin, sliced in a plane parallel to the plane of the skin, is removed from one part of the body and transplanted to the defect somewhere else. With quite elementary surgical care such a graft will heal firmly into place, abolishing the constant loss of body fluids by seepage from the wound and greatly diminishing the risk of infection. The donor area heals without assistance, the surface layer of the skin being restored by upgrowth from the hair roots plus some ingrowth from the margins. Although it requires experience, deftness, and judgment, the surgical procedure is not rated difficult.

In the scenario just outlined, the donor of the graft is also the recipient: it is an autograft. Serious problems arise when the area of skin destroyed is so large that there is not enough sound skin left to provide cover by grafting. The obvious solution—to use an allograft, skin from someone else—is ruled out by the fact that nonself tissues are recognized as such by their recipients and immunologically rejected within a week or two.

The immunologic faculty, like any other, develops and matures over the course of life. It is not present in the embryo, for pieces of embryos can be transplanted from one to another at will: indeed, one embryo may be fused with another to form a genetically composite organism (a chimera) which, having four parents, is referred to as tetraparental—an incompatible union of Greek and Latin.

In surgical practice immunologic incompatibility is counteracted by the use of *immunosuppressive agents.* These are drugs that shut down the immunologic response, especially that arm of it known as cell-mediated immunity, at doses high enough and for periods long enough to allow the graft to be accepted. It is a remarkable dispensation that this procedure should be possible at all. In fact, the use of kidney allografts has now been received into orthodox surgical practice and is no longer considered an experimental procedure; heart and liver transplants are well on the way to achieving the same standing.

The success of clinical transplantation is the outcome of a close and especially fruitful collaboration between laboratory and ward, made possible by the sanguine and adventurous temperament of surgeons combined with their willingness to master—indeed, often to contribute to—the immunologic theory that bears so directly on their work. The limiting factor in clinical transplantation has never been medical prowess, for surgical technique has always been equal to the demands made upon it. Sometimes the physiological barriers are insuperable: the transplantation of the brain, for example, is not a feasible procedure. Nor is much to be expected from grafts of segments of peripheral nerves—but unlike the transplantation of kidney or heart, these are not often matters of mortal urgency.

We should be clear, all the same, that transplantation is by no means confined to organs and solid tissue: the transplantation of populations of cells has been coming rapidly to the fore (blood transfusion is the earliest and best-known example). The benefactions to be expected of cellular engineering include compensating for immunologic deficiencies and defects of blood-forming organs. And it may be by means of cellular engineering that transplantation of hormone-secreting substances will be accomplished.

For a long time the transplantation of organs—particularly the heart—was frowned upon by public figures who arrogated to themselves the responsibility for being the conscience of the world and the spokesmen of a God-fearing, right-thinking minority. These people regarded transplantation as undignified and morally wrong and pleaded instead for natural death. So deep-seated, however, and of such long standing is the human preference for being alive that these spokesmen have not carried their point. We venture to predict that if they themselves were the victims of a disease for which transplantation was the only remedy, they would probably submit to the indignity of being cured.

Vaccination

In the broad sense, vaccination is a procedure for generating an active immunity to an infectious agent. It may be accomplished in alternative ways, both illustrated by active immunization against poliomyelitis. In the original Salk vaccine the immunizing agent (or antigen) was a virulent polio virus rendered harmless by treatment with formaldehyde. In the alternative procedure the subject received active but nonpathogenic virus related closely enough to virulent polio virus to confer protection vicariously (much as infection by cowpox protects against infection by smallpox virus).

Because smallpox was at one time a universal scourge of mankind, prophylactic immunization against it was an even more urgent enterprise than protection against poliomyelitis. Although it was not until the eighteenth century that the procedure came into use in the Western world, variolation was practiced centuries earlier in China and the Near East. The procedure was carefully observed and in due course reported on by the brilliant Lady Mary Wortley Montagu. This attractive young woman (if Sir Godfrey Kneller's portrait is anything to judge by) was poet enough to be sneered at by Alexander Pope and attracted the censure of Samuel Johnson for indecorously preferring Henry Fielding's *Tom Jones* to Samuel Richardson's *Clarissa*. The variolation procedure she advocated involved deliberately infecting a subject with a pustule of a victim suffering from a mild attack of smallpox. It was a safe method that provided effective protection against a serious infection.

It may have been Lady Montagu who aroused the interest of King George I in the procedure. Under his patronage the efficacy of vaccination was tested on several condemned felons (three men and three women) in Newgate Prison. All were inoculated and all survived. Voltaire noted with satisfaction that the prisoners were doubly fortunate, gaining not only their lives (for they were pardoned) but also protection from smallpox. It must be said, however, that his opinion of variolation was that of a literary intellectual rather than of a sage: he clearly thought it droll and characteristically dotty of the English to protect people from smallpox by giving it to them on purpose. James Boswell reports no animadversion to smallpox on the part of that other big gun of the eighteenth century, Samuel Johnson, but he got no nearer smallpox than the fault he found with Lady Montagu's literary judgment.

Because variolation was already widely in use, we cannot credit Edward Jenner (1749–1823) with the introduction of vaccination in the general sense of prophylactic immunization: his great achievement was to turn to good account the observation that persons who had suffered from cowpox enjoyed immunity to smallpox. Jenner accordingly practiced variolation using cowpox rather than smallpox pustules. This is literally *vaccination* (from the Latin *vacca,* a cow) in a very special sense.

Jenner's famous report on his procedure was published in 1798 as "An enquiry into the causes and effects of the *variolae vaccinae,* a disease . . . known by the name of the cow pox." His was a tremendous accomplishment which may be said to have reached its apotheosis when the World Health Organization in 1980 announced the virtual disappearance of smallpox.

Vaccination was introduced into France in 1800 in Boulogne, where the nation's gratitude is marked by a fine iron statue; Edward Boylston of Harvard Medical School who introduced the procedure into Boston is remembered there by a street name. Jenner, despite his impressive discovery, received no such distinction in England. N. McIntyre (*History of Medicine* [March/April 1980]:8) records that Jenner's statue, erected in Trafalgar Square, was judged in the House of Commons to have "no business amongst the naval and military heroes of the country" and would indeed "pollute and desecrate the ground." These were remarks with a Tory bray about them, which could be silenced only by removing the statue to the north side of Kensington Gardens.

Jenner was not made a Fellow of the Royal College of Physicians, though he was elected a Fellow of the Royal Society of London in 1789. An analogous fate befell Jonas Salk, the prime mover in vaccination against poliomyelitis and a person in the very front rank of twentieth-century medical engineers (in the sense of those who translate thought into action). Salk did not receive a Nobel Prize and has not yet been elected to membership in the U.S. National Academy of Sciences—omissions that diminish both honors.

VIRUSES

Measles, poliomyelitis, German measles, hepatitis, and some cancers in experimental animals have in common that they are caused by viruses. Viruses behave as if they were minute living organisms—as if they were bacteria stripped down to the bare essentials of nucleic acid in a protein sheath or capsule. Indeed, the largest viruses such as those giving rise to ornithosis (psittacosis) probably are small bacteria.

It is in many ways injudicious to describe viruses as living organisms; rather, they are infective agents which have the power of subverting their hosts in such a way that replicas of themselves are formed. They are not themselves capable of multiplication by division, nor can they live outside the environment provided by a living cell. At the time of this writing no known antibiotic acts specifically upon a virus; control of virus infection is essentially a matter of prophylaxis by immunization with living attenuated viruses related antigenically to the virus against which protection is sought, or with virulent virus killed by treatment with formaldehyde.

Inasmuch as viruses are made known only by their causing disease or other pathological changes, the existence of benign viruses having no ill effects remains conjectural. No virus is *known* to do good: it has been well said that a virus is "a piece of bad news wrapped up in protein."

VITALISM

This is a doctrine that takes various forms, which have in common a flat repudiation of the idea that a living organism's vivacity*—its state of being alive—can be explained satisfactorily in terms of its form and composition; that is, in terms of what it is made of and how those constituents interact physically and chemically. Some immaterial vital principle is required in addition. For Hans Driesch, it was an entelechy; for Henri Bergson (George Bernard Shaw concurring), an *élan vital*.

Driesch's position is the more straightforwardly Aristotelian. The train of thought from which it arose can be inferred from the experiments described by Driesch in his famous Gifford Lectures of 1907 (*The Science and Philosophy of the Organism* [London, 1908]). The relevant embryological experiments are these: in the normal course of events when a fertilized egg divides in two, the clone or descendants by successive cell divisions of each daughter cell gives rise to half the body. No vast surprise was generated, therefore, when the pioneer of experimental embryology Wilhelm Roux (1850–1924) killed one of the two daughter cells resulting from the first division of a fertilized sea urchin's egg and found that a half-egg gave rise to a half-embryo. Results similar in principle were obtained using the egg of a frog, of a mollusk, and of a jelly-like pelagic organism belonging to the phylum *Ctenophora*. All this was very much what was to be expected; Driesch, however, performed the same experiment in another way on another sea urchin. His results were quite different: either one of the first two daughter cells into which the fertilized egg divided gave rise to a small but *whole* embryo—and the same applied to any one of the four daughter cells produced by the second cleavage of the fertilized egg. Indeed, even the division into two of a fairly advanced embryo resulted in the formation of two whole organisms.

From these experiments Driesch inferred (and every subsequent experiment has upheld him) that differentiation in development cannot be due to a succession of unequal nuclear divisions, that is, to unequal partition of the cell substance during successive cleavages of the fertilized egg. More than that, it convinced Driesch that the organism could not be interpreted as if it were behaving like a machine, however subtle or exquisite its working. In the process of "regulation," whereby that which was potential became actual, Driesch clearly saw the embodiment (if

* An attractive seventeenth-century designation for that which we nowadays pallidly refer to as "vitality."

such a word can be used of a force or spirit) of Aristotle's entelecheia. So Driesch's vitalistic or autonomous factor that governed morphogenesis was named entelechy as a mark of admiration for Aristotle's genius.

Where stands vitalism today? It is virtually impossible to think of any observation or experiment, except perhaps the total synthesis of some living organism, that could falsify the notion of vitalism or an inference drawn from it. Until this has been done, as one day it may well be, the concept of vital force or *élan vital* must be judged outside science. But no one strikes attitudes about the matter any more: vitalism is in the limbo of that which is disregarded. Modern biologists do not find it necessary to appeal to a disembodied vital principle such as Driesch envisaged. The state of being alive seems to us to be best described as an emergent property, in the sense conceived in the entry REDUCTIONISM. Reading Driesch's lectures does, however, remind us how much there is still to explain and how far we still are from a full theoretical understanding of development.

VITAMINS

No substantive property distinguishes a vitamin; it is any material essential for the growth or continued health of an organism that does not happen to be synthesized by the body, so that it must form part of the diet. One animal's vitamin is not necessarily another's: for us, but not for mice, ascorbic acid is a vitamin. Substances that are vitamins for higher animals must be widespread and fairly abundant, otherwise the animals known to require them would not be alive to bear witness; thus vitamin C and the vitamins of the B complex are broadly distributed in the diets of the animals that require them.

The notion that teleology of vitamins is to prevent a variety of metabolic diseases has been highly misleading: it has created the impression that vitamins need only be consumed in quantities sufficient to prevent the disease which their insufficiency would instigate. Opinion is changing, however, and there is growing evidence that people need more vitamin C than the relatively small quantity necessary and sufficient to prevent scurvy. *Mutatis mu-*

tandis, the same may be true of vitamin A, which epidemiologic research has shown to play an important part in resistance to cancer. Some vitamins (the fat-soluble ones) are harmful in excess, and the enthusiastic consumption of vitamin tablets as a kind of panacea for most human ailments cannot be unreservedly commended.

WATER

Land animals are about three-fourths water. ("Even the Pope is 70 percent water," said J. B. S. Haldane, characteristically indifferent to the sensibilities of the Catholic community.) But of course water has other claims on the attention of biologists: in development, water is the cradle of life, as in evolution it was the nursery.

Even those who know nothing else of science know that water has an atomic structure represented by the empirical formula H_2O. We hope the world will not come tumbling about their ears when they learn that water exists in forms with a molecular structure more accurately rendered by the empirical formulas H_4O_2, or even H_8O_4. The physical properties of water sometimes take extreme values that interest biologists especially. For instance, it takes more heat to raise the temperature of a unit volume of water through 1° C than is needed for any other substance; this property greatly facilitates the distribution of heat in the body and the transport of heat by ocean currents (such as those of the Gulf Stream and the North Atlantic).

The latent heat of evaporation of water (measured in terms of the heat energy required to cause the evaporation of 1 ml) is higher than that of any other substance. This too has geophysical and biological implications. Biologically, it provides for the regulation of body temperature by evaporation; geophysically, it provides for the maintenance of a more equable temperature on the earth, because the vast amount of solar heat absorbed upon evaporation of the surface waters of the ocean is repaid during the condensation that precedes the fall of rain.

Another curious physical property of water is that it is denser at 4° C than as ice at 0° C, with the result that ice floats and that large bodies of water freeze only at the top. A lake will not freeze solid unless freezing extends right from the top to the bottom layers, something which the rather low thermal conductivity of water makes unlikely. Just as well, for a lake frozen solid would not thaw out the whole year through.

These biologically felicitous properties of water were an important element of Lawrence Joseph Henderson's (1878–1942) interesting but philosophically unsound variant of the argument from design. His central point in *Fitness of the Environment* (New York, 1913) was that far from being random and fortuitous in origin, the natural environment provided by the earth is so well suited to the emergence of life that we may think of it as "preadapted" for such a phenomenon. The argument is shaky because properties of living things (which include the ability to make proposals such as this one) are what they are because properties of the surrounding world were what they were. We cannot look back on them and say retrospectively how fortunate it is that they were so, because there is no conceivable alternative.

The human body loses water in the urine and perspiration, and by evaporation from the surface of the body and lungs; but, like other mammals, it conserves salts very efficiently. Although salt passes through the filtering element of the kidney, it is reabsorbed into the body. Water as water (meaning water not already preoccupied as a solvent of salts and soluble body constituents) must therefore be taken in to reinstate the appropriate concentration of body fluids.

The comparative ease with which civil engineering has made it possible for human beings to quench their thirst has dulled our sensibilities to the predicament of seals and whales: their body fluids have much the same composition as our own and their need for water is much the same. Like the Ancient Mariner, although surrounded by water, they cannot drink; for seawater has a higher concentration of salt than their own body fluids, and to drink it merely exacerbates their need. The sea is arid for mammals. Fresh water is something of a gourmet delicacy for whales and seals, and when seals have access to fresh water, they drink with what is known as "evident relish." Unhappily, such good fortune does not often come their way. Their requirement for free water, though somewhat diminished by the excretion of a more concentrated urine than that of land mammals, is aggravated by the loss of water inevitable in nursing whales and seals (even though the milk of whales is relatively dense and buttery). Somehow the water balance is righted by a gain of water from the combustion of carbohydrate and especially of fat—processes that liberate "water of metabolism"—and from body fluids of the invertebrates that constitute so high a proportion of the diets of whalebone whales. The metabolism of organic matter in these body fluids eventually frees the water to play an osmotically active role in the whale's body, the whalebone filters removing much of the seawater that might otherwise be taken in with the food.

One cannot slake thirst by drinking a fluid with the same salt concentration as one's own body fluids. The Ancient Mariner cannot be reproached, however, for biting his arm to suck his blood; his purpose was not to quench his thirst, but to wet his whistle to cry "A sail, a sail!"—an exclamation he might not otherwise have been able to pass.

WOODGER'S PARADOX

It is the defining characteristic of a paradox that strictly logical reasoning from accepted premises may lead to inferences repugnant to common sense. A paradox, it has been accurately said, has the same significance for a philosopher or logician that the smell of burning rubber has for an electronics engineer: it is a signal that something is amiss.

Joseph Woodger, distinguished mathematical logician and biologist and author of *Biological Principles* (London, 1929) and *The Axiom Mathematic Method in Biology* (Cambridge, 1937), circulated among his friends, for their amusement, the paradox outlined below.

Consider any taxon: we shall choose *phylum,* though class, genus, or species would do as well. Let it be a taxonomic principle that every organism must belong to some one phylum and no organism may belong to two. If this is so, Woodger reasoned, the evolutionary step between one phylum and the next must take place between one generation and the next, that is, between a parental and a filial generation. If an organism belonging to phylum A evolves into an organism belonging to phylum C, there can be no intermediate form B that is a member of both A and C nor, alternatively, can it be that organism B is in a taxonomic limbo—a member of no phylum—for both suppositions flout the taxonomic canons agreed upon above. There is no alternative, then, but to suppose that A as a member of one phylum must have evolved directly into an organism C that belongs to another phylum. Thus the progression from one phylum to the next, being without intermediates, must take place from one generation to the next.

What is wrong with this reasoning?

Woodger, a rather old-fashioned biologist, envisaged evolution in what even in his own lifetime became a rather old-fashioned way—in the genealogical or family-tree style. It is *this* that is at fault. In the discussions on population genetics, gene frequency, neo-Darwinism, and the paragraphs in the entry TAXON that deal with species notation, it is made clear that the genealogical or family-tree conception of evolution is incompatible with population genetics: it is not one organism that evolves into another, but one population characterized by some distinctive pattern of gene frequency that evolves into another population. Even the notion of species membership is shown to be dubious from the population-genetic point of view.

Our choice of phylum as the taxon in the above example was made to give Woodger's paradox special rhetorical force. *Species* would no doubt have been a better choice; had species been chosen, we should have been led to the equally paradoxical view that a step from one species to another also occurs in one generation, a conception equally inadmissible from the population-genetic point of view. It is the formulation of such antinomies as Woodger's that leads us to undertake the anxious self-scrutiny on which the long-term welfare of a science so largely depends. The older evolutionists, secure in the genealogical conception of evolution, would have been plagued by no such worries as these and might indeed have dismissed them as illusory.

WORM

Here is a word that is a good example of the kind of designation taxonomists dismiss as trivial. Not even the seven alternative spellings recognized by the *Oxford English Dictionary* (worm, wyrme, wirm, wrim, wrm, wurm, wormr) are inducement enough to take the word seriously in the biological sense, for taxonomists are adamant that all worms properly so called are members of the phylum Annelida. This classification of course includes the common earthworm ("whose modest unobtrusiveness and busy beneficence are a lesson to us all") but as constituted does not include flatworms, roundworms, hookworms, tongue worms, mealworms, arrowworms, acornworms, peanut worms, eelworms, echinoid worms, tapeworms, velvet worms, beard worms, or thorny-headed worms, all of which live with the odium of being improperly named.

The Annelida constitute a well-defined phylum of about ten thousand species of tubular animals, built to a segmental pattern with regularly spaced annular grooves visible from the outside. The segments have a similar plan, each with a nerve ganglion, an excretory organ, a number of bristles that provide for attachment to the substratum, and a disposition of muscles that makes possible the characteristic wormlike movement—extension of the front end, followed by drawing up of the rear. There is a spacious body cavity and the blood is contained within special blood vessels (the vascular system is not—as it is in arthropods—of the open type). The most abundant annelids are the so-called bristle worms, and the group also includes leeches.

Zoos

There can be few bright teenagers who, looking first at a cage of monkeys and then at the spectators gaping at them, have not had dark and (as it has seemed to them) rather deep thoughts upon the theme of which group should be regarded as exhibits and which as lookers-on.

Zoos and menageries are entertainments for all cultures, at all times and in all places—full of much that is edifying, diverting, and, for reasons hard to put into words, moving.

The authors of this volume, zoologists by education and inclination, have many recollections of wonders witnessed at the zoo: among them a compound of giraffes which, though incessantly and vigorously in motion, weaving this way and that way in complete silence, never bumped into one another or even touched. We remember the delight (it would be behaviorist pedantry to describe it otherwise) of elephants being hosed down with cold water on a very hot day. We remember too, because it showed up human beings in such a poor light, the public copulation of two turtles—a majestic performance at the unhurried tempo of the Mesozoic, not visibly affected by the vulgar laughter and camera clicking that surrounded them. Moving? Yes, perhaps because turtles and reptiles generally have had their day, so what have their young to look forward to?

Such entertainment is all very well, but is it right that animals should be captured, confined, and exhibited for the amusement or instruction of human beings? *Of course not!* But consider: this is an imperfect world, and it does not seem to us that nonhuman animals are disproportionately ill used in view of the fact that so many human beings live in dark, unhygienic shacks or shanties that are in some ways inferior to the quarters allotted to animals in good modern zoos. Happily, there are signs of better things on the way, especially game parks, nature preserves, and other approximations to a natural habitat. This is still only a beginning and there is plenty of room for magnanimity by U.S. tycoons with conscience money to expend or by dotty British noblemen whose estates may as well be possessed by predators of their own choosing rather than by Her Majesty's Collectors of Taxes. Besides, let it not be forgotten that some species of animals are with us today only because zoos offer them homes.

INDEX

Abercrombie, Michael, 61. *See also* Contact inhibition
Abortions: clinical, *see* Feticide; spontaneous, 15
Acetic acid, 12
Acetylcholine, 22–23, 192. *See also* Anticholinesterases
Achondroplasia, 18
ACTH, *see* Adrenocorticotrophin
Active immunity, 153
Act of Creation (Koestler), 142
Actuarial measurement of aging, 5–6
Adaptation: sensory, 1, 235; irritability and, 164, 235; mimicry as, 184–185, 193. *See also* Congenital disorders; Evolution and evolutionary theory; Fitness; Immunologic processes; Nature, errors of
Addison's disease, 152
Adenovirus-*12,* 187, 220
Adrenal gland, 2–3, 65, 76, 77–78, 79–80
Adrenaline, 2
Adrenocorticotrophin (ACTH), 3
Africa, 36
After Many a Summer (Huxley), 200
Agammaglobulinemia, 24
Age distributions, 3–4
Age-specific death rates, 68, 101–102, 172
Aggression, 4, 147; toward animals, 19–20; quasi-ethological interpretations, 85
Aging, *see* Growth, laws of biological; Senescence
Agriculture, 25–26
Albinism, 8–9, 111
Aldosterone, 2–3
Algae, 247
Alice in Wonderland (Carroll), 67

Allantois, 110, 122–123
Alleles, 117; defined, 110–111; and dominance and recessiveness, 112–113; mutation and, 120
Allelomorphic, defined, 111
Allergies, 9–10, 24, 36, 80, 155
Allografts, 122, 123, 154, 156, 270–271
Allometric growth, 10–11
Altruism, 11–12, 171, 240. *See also* Sociobiology
Amateurs, in science and philosophy, 74, 120, 144, 150, 240–241
Ambystoma, 199
Amino acids, 12, 92, 113, 194–195, 209
Ammonia, 74
Amniocentesis, 72, 111, 125
Amnion, 111
Amoebas, 222, 258
Amphibians, 119, 121, 252, 254
Amphioxus, 13–14, 164, 225
Anabiosis, 14, 262–263
Anabolism, 3
Analogy, 145
Anaphylactic shock, 10, 80
Anaplasia, 47–48
Anatomy, comparative, 40, 60–61, 145
Anencephaly, 14–15, 244
Animal husbandry, 25–26
Animals: stewardship and, 15–20, 191, 284–285; experimentation on, 16–18, 79–80, 82–83, 84; Aristotle on, 27–28; behavior, *see* Ethology; geographic distribution, 132; in zoos, 284–285. *See also* Stockbreeding
Annelids, 30, 163, 248, 283
Anseriformes, 253
Anteaters, 185, 191; adaptation in, 1
Anthropomorphs, 24–25
Antibiotics, 20–22, 251, 263; allergic re-